Lecture Notes in Networks and Systems 973

The series "Lecture Notes in Networks and Systems" publishes the latest developments in Networks and Systems—quickly, informally and with high quality. Original research reported in proceedings and post-proceedings represents the core of LNNS.

Volumes published in LNNS embrace all aspects and subfields of, as well as new challenges in, Networks and Systems.

The series contains proceedings and edited volumes in systems and networks, spanning the areas of Cyber-Physical Systems, Autonomous Systems, Sensor Networks, Control Systems, Energy Systems, Automotive Systems, Biological Systems, Vehicular Networking and Connected Vehicles, Aerospace Systems, Automation, Manufacturing, Smart Grids, Nonlinear Systems, Power Systems, Robotics, Social Systems, Economic Systems and other. Of particular value to both the contributors and the readership are the short publication timeframe and the world-wide distribution and exposure which enable both a wide and rapid dissemination of research output.

The series covers the theory, applications, and perspectives on the state of the art and future developments relevant to systems and networks, decision making, control, complex processes and related areas, as embedded in the fields of interdisciplinary and applied sciences, engineering, computer science, physics, economics, social, and life sciences, as well as the paradigms and methodologies behind them.

Indexed by SCOPUS, INSPEC, WTI Frankfurt eG, zbMATH, SCImago.

All books published in the series are submitted for consideration in Web of Science.

For proposals from Asia please contact Aninda Bose (aninda.bose@springer.com).

Phayung Meesad · Sunantha Sodsee ·
Watchareewan Jitsakul · Sakchai Tangwannawit
Editors

Proceedings of the 20th International Conference on Computing and Information Technology (IC2IT 2024)

Editors
Phayung Meesad
Faculty of Information Technology
and Digital Innovation
King Mongkut's University of Technology
North Bangkok
Bangkok, Thailand

Sunantha Sodsee
Faculty of Information Technology
and Digital Innovation
King Mongkut's University of Technology
North Bangkok
Bangkok, Thailand

Watchareewan Jitsakul
Faculty of Information Technology
and Digital Innovation
King Mongkut's University of Technology
North Bangkok
Bangkok, Thailand

Sakchai Tangwannawit
Faculty of Information Technology
and Digital Innovation
King Mongkut's University of Technology
North Bangkok
Bangkok, Thailand

ISSN 2367-3370 ISSN 2367-3389 (electronic)
Lecture Notes in Networks and Systems
ISBN 978-3-031-58560-9 ISBN 978-3-031-58561-6 (eBook)
https://doi.org/10.1007/978-3-031-58561-6

This Springer imprint is published by the registered company Springer Nature Switzerland AG
The registered company address is: Gewerbestrasse 11, 6330 Cham, Switzerland

Preface

The rapid evolution of technology has profoundly impacted society, driving digital innovation across various sectors. In today's global market, the demand for skilled IT workers is paramount. Their expertise in cutting-edge technologies is essential for driving efficiency, productivity, and competitiveness across industries.

In delving into the contents of this book, it is imperative to recognize the pivotal role played by IT professionals in molding the future digital landscape.

As a platform for knowledge exchange and innovation, the International Conference on Computing and Information Technology (IC2IT) has been at the forefront of facilitating collaboration and driving digital advancements for over twenty years. In continuation of this mission, this book presents the key contributions of researchers who participated in the 20th International Conference on Computing and Information Technology (IC2IT 2024), held from May 16 to 17, 2024, in Bangkok, Thailand.

The conference proceedings feature meticulously selected and peer-reviewed research papers spanning various domains, including machine learning, natural language processing, image processing, robotics, and digital technology. Submissions from six countries underwent rigorous evaluation by at least three members of the program committee, resulting in the acceptance of 15 papers for publication.

Through this compilation, our aim is to provide readers with comprehensive insights into the latest advancements in digital technology and their profound impact on society. We are confident that this resource will prove invaluable to researchers, practitioners, and enthusiasts alike, fostering further digital innovation and progress.

Finally we extend our heartfelt gratitude to the authors, program committee members, and collaborating institutions in Thailand and abroad for their unwavering support and academic collaboration. As well as special thanks are also due to the dedicated staff members of the Faculty of Information Technology and Digital Innovation at King Mongkut's University of Technology North Bangkok.

February 2024

Phayung Meesad
Sunantha Sodsee
Watchareewan Jitsakul
Sakchai Tangwannawit

Organization

Program Committee

M. Aiello	UNI-Stuttgart, Germany
S. Auwatanamongkol	NIDA, Thailand
T. Bernard	Li-Parad, France
N. Bhumpenpein	KMUTNB, Thailand
S. Boonkrong	SUT, Thailand
N.H.H. Cuong	UD, Vietnam
T. Eggendorfer	HS Weingarten, Germany
S. Hengpraprohm	PBRU, Thailand
K. Hengproprohm	PBRU, Thailand
D.V. Hieu	TGU, Vietnam
S. Hiranpongsin	UBU, Thailand
W. Janratchakool	RMUTT, Thailand
T. Jensuttiwetchakul	KMUTNB, Thailand
W. Jitsakul	KMUTNB, Thailand
T. Kaewkiriya	PIM, Thailand
S. Krootjohn	KMUTNB, Thailand
P. Kropf	Uni NE, Switzerland
P. Kunakornvong	RMUTT, Thailand
U. Lechner	UniBW, Germany
N. Lertchuwongsa	PSU, Thailand
M. Maliyaem	KMUTNB, Thailand
P. Meesad	KMUTNB, Thailand
K. Pasupa	KMITL, Thailand
K. Saengkrajang	PCRU, Thailand
P. Saengsiri	TISTR, Thailand
T. Sarawong	RMUTK, Thailand
S. Smanchat	KMUTNB, Thailand
S. Sodsee	KMUTNB, Thailand
G. Somprasertsri	MSU, Thailand
W. Sriurai	UBU, Thailand
T. Sucontphunt	NIDA, Thailand
P. Tangwannawit	PCRU, Thailand
S. Tangwannawit	KMUTNB, Thailand
K. Tantayakul	PSU, Thailand
J. Tengsuthiwat	KMUTNB, Thailand

C. Thaenchaikun	PSU, Thailand
J. Thaenthong	PSU, Thailand
J. Thongkam	MSU, Thailand
N. Tongtep	PSU, Thailand
K. Treeprapin	UBU, Thailand
N. Utakrit	KMUTNB, Thailand
N. Utakrit	KMUTNB, Thailand
K. Viriyapant	KMUTNB, Thailand
V. Wattanasoontorn	PSU, Thailand
P. Wongchampa	KMUTNB, Thailand

Organizing Partners

In Cooperation with

King Mongkut's University of Technology North Bangkok (KMUTNB)
Council of IT Deans of Thailand (CITT)
FernUniversität in Hagen, Germany (FernUni)
Chemnitz University of Technology, Germany (CUT)
Oklahoma State University, USA (OSU)
Edith Cowan University, Western Australia (ECU)
Hanoi National University of Education, Vietnam (HNUE)
Mahasarakham University (MSU)
National Institute of Development Administration (NIDA)
Rajamangala University of Technology Thanyaburi (RMUTT)
Kanchanaburi Rajabhat University (KRU)
Ubon Ratchathani University (UBU)
Gesellschaft für Informatik (GI)
IEEE CIS Thailand

Contents

Identifying Key Issues to Enhance the Cybersecurity Awareness Strategy within Organizations

Anawin Kaewsa-ard and Nattavee Utakrit(✉)

Faculty of Information Technology and Digital Innovation, King Mongkut's University of Technology North Bangkok, Bangkok, Thailand
s6307021910041@email.kmutnb.ac.th, nattavee.u@itd.kmutnb.ac.th

Abstract. The impact of cybersecurity threats has had a significant negative effect on organizations in terms of financial, operational, and external stakeholder trust. Promoting awareness of the importance of secure technology usage in the cyber world is crucial. This research aims to identify key issues affecting the establishment of cybersecurity awareness within organizations and analyze qualitative data from the perspective of cybersecurity experts through in-depth interviews with 15 participants. The interview questions were systematically developed to validate their ability to explain phenomena affecting cybersecurity awareness. Analysis of the qualitative data revealed four significant issues that influence cybersecurity awareness within organizations including security risks faced by organizations, cyberattacks on individuals, threats to organizational security and problem-solving guidelines. In conclusion. The research findings highlight the importance of fostering cybersecurity awareness for all individuals within an organization and consider it a vital component in sustaining cyber security.

Keywords: Security Awareness · Qualitative Analysis · In-depth Interview · Thematic Analysis

1 Introduction

The global cybersecurity research institution has compiled a report on cyber risk that has garnered significant attention and concern from top-level executives worldwide. The top three risks include: 1) Phishing attacks, which deceive individuals through cyber means such as email, SMS, and vishing. 2) Business email compromise involving identity spoofing within organizations 3) Ransomware attacks [1]. Furthermore, the report discusses challenges in managing cybersecurity within organizations including 1) budget constraints 2) inadequate employee training 3) cybersecurity staff shortages 4) a lack of executive support 5) communication issues 6) organizational culture 7) insufficient cybersecurity skills 8) limited training time and 9) a lack of cybersecurity investment return assessment [2]. Sustainable solutions to these issues involve establishing a dedicated cybersecurity awareness and training department within organizations as well as ongoing evaluation of internal cybersecurity awareness effectiveness. It is evident

P. Meesad et al. (Eds.): IC2IT 2024, LNNS 973, pp. 1–11, 2024.
https://doi.org/10.1007/978-3-031-58561-6_1

that technology alone cannot entirely mitigate cyber threats as it is influenced by various factors including people, processes, and technology. People play a critical role in cybersecurity and are affected by a lack of awareness and self-protection skills [3–5]. Therefore, maintaining effective cybersecurity within organizations remain challenging. Organizations should prioritize fostering cybersecurity awareness among all individuals to protect their data in the digital age. This underscores the need for a sustainable approach to enhancing cyber security awareness within organizations through qualitative research. Such an approach will promote secure technology usage and sustainable cybersecurity practices in the future.

2 Related Work

2.1 Cyber Security Awareness

Awareness is defined in NIST Special Publication 800-16 as follows Awareness is not training. Security awareness efforts are designed to change behavior or reinforce good security practices. The purpose of awareness presentations is simply to focus attention on security. Awareness presentations are intended to allow individuals to recognize IT security concerns and respond accordingly. In awareness activities, the learner is the recipient of information whereas the learner in a training environment has a more active role. Awareness relies on reaching broad audiences with attractive packaging techniques. Training is more formal, having a goal of building knowledge and skills to facilitate job performance [6].

The cybersecurity problem arising from the lack of organizational-level awareness results from computer users not possessing sufficient knowledge of cybersecurity. This lack of awareness makes users vulnerable targets for hackers and creates vulnerabilities that can be exploited for cyberattacks. Addressing these issues requires proper training to impart the skills necessary to deal with cyber threats effectively. Training should shift from mere lectures to interactive scenarios and response strategies. By applying practical training in real life situations, individuals can genuinely incorporate the knowledge gained from cybersecurity training into their daily lives [7].

Our review of current literature highlights those cyber-attacks targeting user awareness often exploits human vulnerabilities rather than relying solely on advanced technical expertise. Despite this, many organizations prioritize investment in sophisticated technological defenses to combat a variety of threats. However, this approach overlooks the fact that employees frequently utilize email, social media, and other online platforms as part of their daily work. Therefore, effective prevention of cyberattacks cannot be achieved through technology alone and understanding the human element is essential [8]. Human centric cyberattacks involve the interaction between the attacker and the victim who typically possesses valuable information. Attackers gather sensitive data from their targets and then use this information to establish trust and build a relationship. Despite advancements in technology for threat prevention, insufficient attention has been given to addressing human factors. The researcher has compiled a summary of studies concerning global cyber knowledge, including research findings. Additional insights into problem solving are presented in Table 1.

In the Thailand security awareness behavior study, Daengsi et al. conducted research on the cybersecurity awareness of Thai employees within the financial and banking sectors, with a sample size of 20,134 individuals. This research consisted of three phases: Phase 1, which involved phishing mail attacks; Phase 2, focusing on knowledge dissemination through e-learning and awareness activities related to phishing mail; and Phase 3, centered around creating cybersecurity situations via a fake promotion campaign with phishing mail attacks. The results of the study revealed a statistically significant increase in the cybersecurity awareness of the Thai employee sample (P-values < 0.05), leading to a 71.5% decrease in the number of Thai employees who opened phishing mail. Additionally, it was observed that female Thai employees displayed higher cybersecurity awareness compared to their male counterparts, and there was no significant difference in cybersecurity awareness among different generational groups, including Gen X, Gen Y and Baby Boomers. Consequently, fostering sustainable cybersecurity awareness among employees within organizations is influenced by factors such as age and role [9]. This research demonstrates that imparting knowledge in security awareness genuinely influences users' behaviors in coping with cyber threats.

2.2 Qualitative Analysis

The analysis of qualitative data from various sources, such as interviews, group discussions, and documents, aims to develop theories based on grounded evidence found in the data's source. The process of theoretical making, following the grounded theory approach, doesn't involve predefining practical agreements. Qualitative data analysis can be conducted within a context, considering aspects originating from the data sources, including processes, philosophies, and literature as well as methods and practical guidelines. The researcher objective is not necessarily to construct a new theory but rather to comprehend the phenomena emerging from the grounded theory [10]. The grounded theory analysis technique was employed to analyses the data. A coding process consisting of open axial and selective coding was used to identify themes related to key issues of security awareness problem [11]. Qualitative data analysis has been widely applied in research due to its systematic conceptual approach, which aids in uncovering relationships within data, interpreting meaning appropriately, and considering the perspectives and attitudes of the interviewees when investigating emerging phenomena [12].

Erlingsson and Brysiewicz introduced a systematic approach to qualitative data analysis through the thematic analysis process. This approach involves initially collecting data from interviews with a sample group. Subsequently, the data is organized into coherent sentence patterns that convey meaning clearly and comprehensibly. This organization of data is carried out to facilitate the explanation of whether the observed phenomena align with any existing theories or not [13].

3 Methodology

Research Questions

RQ1: How does the cultivation of security awareness impact organizational dynamics?
RQ2:How does fostering cyber awareness behavior serve as a foundational approach to addressing user-related cyber issues?

3.1 Participant's Information Criteria and Recruitment

The researcher carefully selected a group of 15 cybersecurity experts who hold positions at both operational and managerial levels [14]. The criteria for selecting these experts included considerations of their educational background, job roles, and professional experience. Specifically, the chosen individuals were required to possess a master's degree in computer science, information technology, or a closely related field. Additionally, they were expected to hold a cyber security certification accredited by ANSI, ensuring compliance with the ISO 17024 standard, thereby signifying their recognized expertise in the field [15]. Furthermore, these experts were mandated to have significant experience with a minimum of 15 years in cybersecurity management or at least 5 years in cybersecurity operations. This stringent selection process aimed to ensure that the participants possessed substantial knowledge and practical insights into cybersecurity issues. By employing this meticulous approach, the research aimed to conduct a comprehensive examination of cybersecurity maintenance concerns across various organizational levels. This endeavor sought to shed light on the diverse perspectives and challenges encountered within the cybersecurity domain with the ultimate goal of enriching the understanding of these issues through face-to-face interviews [16, 17].

3.2 Semi-structured Interview Development

The open-ended questions in the interview questionnaire were developed using the Working Backward Technique in Fig. 1 as identified in prior research [18]. This method was chosen after studying the literature, which showed that designing open-ended questions enables volunteers to respond from various perspectives thus avoiding the potential limitations of researcher-defined questions. The primary advantage of open-ended questions in interviews is that they allow volunteers to explain the causes and consequences of their responses. This approach facilitates the comparison of concordance among volunteers, helping to determine whether their attitudes align in the same direction.

In order to confirm that the questions could genuinely explain the phenomenon affecting the creation of cyber awareness. The researcher conducted a literature review related to the development of cyber awareness behavior. This was done to comprehensively study the issues and solutions systematically, with details presented in Table 1 as follows:

Based on the findings obtained from the research reviewed, researchers recognized challenges linked to insufficient cyber awareness. This realization prompted the formulation of interview questions tailored to engage experts, with support from the working backward steps technique. Consequently, this approach significantly improved the examination of these identified issues.

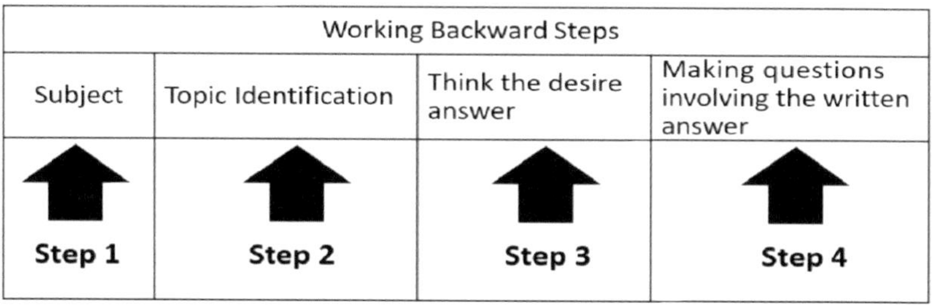

Working Backward Steps			
Subject	Topic Identification	Think the desire answer	Making questions involving the written answer
Step 1	Step 2	Step 3	Step 4

Fig. 1. Working backward process.

3.3 Data Analysis

The researcher constructed a framework derived from the data using Grounded Theory [23, 24]. Employing a coding process involving three distinct types. Open Coding to categorize data, Axial Coding to establish relationships among coded data and Selective Coding for synthesizing the data into a linguistic structure [17, 25, 26]. The resulting framework is presented as a Thematic Map. The analysis was conducted using QDA Miner Lite 3.0, a software package designed for qualitative data analysis to identify patterns in the thematic map, as depicted in Fig. 2. QDA Miner played a crucial role in examining coding frequency and the predominant attitudes among the 15 interviewed experts, represented as a percentage of cases (%Cases). Moreover, it assisted in organizing codes into distinct categories, thereby contributing to a comprehensive understanding of the observed phenomena.

4 Results and Discussion

Through qualitative data analysis obtained from expert interviews using grounded theory, it was found that there are 20 codes that can be categorized into four groups. Each of these groups represents critical factors influencing cyber security awareness in organizations including 1) security risks organizations are facing; 2) cyberattacks on individuals; 3) Threats to cyber security in organization and 4) problem-solving guidelines. Further details can be observed in the Thematic Map in Fig. 3 along with the frequency distribution of the codes resulting from expert feedback as illustrated in Fig. 4.

From Fig. 4, it is evident that the most frequently mentioned code. Ranked first is Building a security behavior with a frequency of 60 occurrences. Second in order is Management Support with a frequency of 48 occurrences. Thirdly, Issues of User Cyber Awareness is highlighted, with a frequency of 44 occurrences. Importantly, all experts concur in the same direction that promoting cyber awareness behavior among individuals within the organization and addressing the issue of user awareness deficiency have far reached implications not limited solely to individual users but extending to operational activities and the organization's information systems. Consequently, it is advisable for the management to formulate strategies or plans for cultivating cyber awareness. To provide a comprehensive overview of the identified key issues in this study, the researchers have organized the codes and expert feedback as exemplified in Table 2. And have summarized the most frequently mentioned codes by experts in each area, as detailed in Table 3.

Table 1. Summary of reviewed sources.

Authors	Purpose	Major Themes	Finding
Jonathan and Kevin (2018) [19]	To determine the IS awareness and behaviors	Security Awareness Behavior	Comprehensive security awareness training will be essential for institutions of higher education as a means of minimizing threats to information technology resources
Nourah et. al. [20]	Study the pattern security awareness and cyber-crime in Saudi Arabia	Measure Awareness of Cybercrime	Most participants need more education and content enrichment with cybercrime and how to deal with it according to local legislation in Saudi Arabia
Tanja and Dietmar (2021) [21]	study both individual and organizational factors that affect IS awareness	Intention to resist social engineering	Information security awareness was confirmed as a central factor for IS
Yan, Ram and Kuang (2014) [22]	To evaluate the influences of key components of comprehensive IS programs on security culture	Building an IS culture	The results indicate that Security Education Training Awareness (SETA) programs awareness has significant influences on security culture and on employee awareness of organizational security policy, and that the awareness of security monitoring also impacts security culture

*IS = Information Security

From the research question, it can be concluded that all experts recognize the importance of fostering cyber awareness behavior for everyone within the organization, as it plays a role in maintaining cybersecurity. By instilling these behaviors through training and development processes, it is possible to transform personnel, including potential insider threats into active contributors to the organization's cyber threat defense. Additionally, there is a consensus that security awareness is a crucial factor in safeguarding

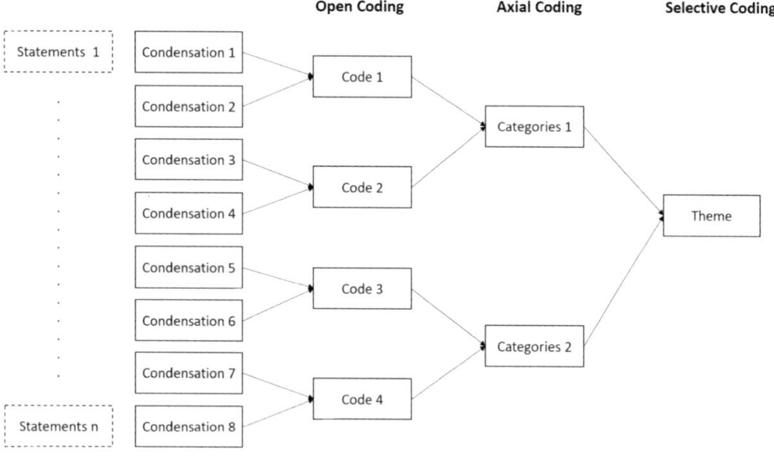

Fig. 2. Thematic map displaying.

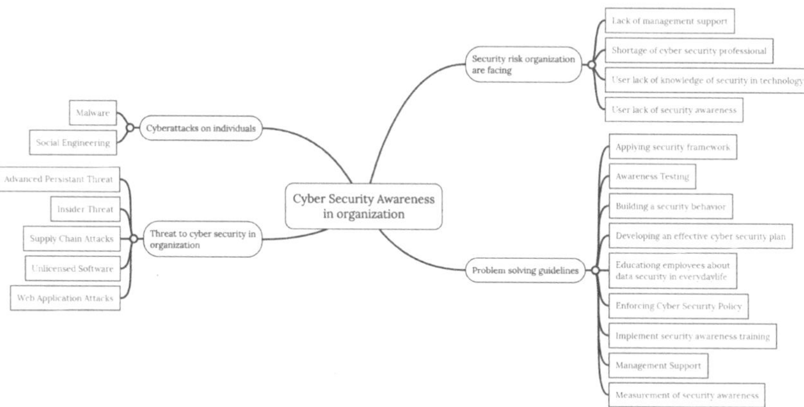

Fig. 3. Finalized thematic map.

cyber security and reducing the potential impact of threats. Furthermore, it contributes to the establishment of a sustainable culture of security within the organization.

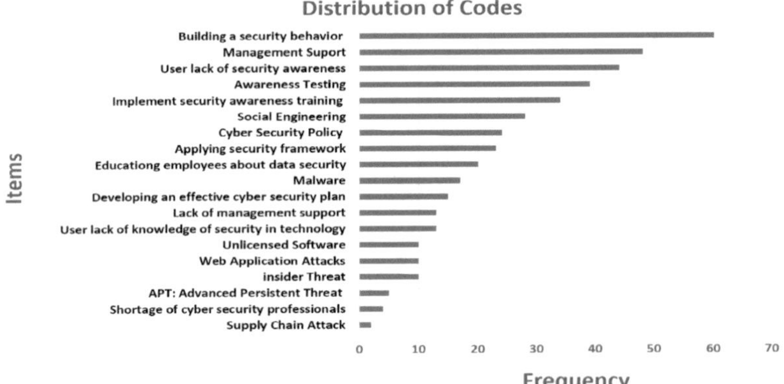

Fig. 4. Coding frequency.

Table 2. Example of coding scheme.

Categories	Code	Data Extracts
Cyberattacks on individuals	Social Engineering	One of the most encountered cyber threats in our daily lives today is phishing attacks. These attacks typically involve enticing individuals to click on a link that leads to a deceptive advertisement or an unfamiliar website often encouraging them to enter personal information. This matter is frequently discussed among students and colleagues in the workplace
Threat to cyber security in organization	Web Application Attack	Commonly encountered organizational level cyber threats often manifest as web-based attacks such as cross-site scripting (XSS), SQL injection and web defacement
Problem solving guidelines	Building a security behavior	I believe that fostering a security culture requires every department within the organization to prioritize the cultivation of cyber awareness. It is not solely the responsibility of the cyber security team, as cyber awareness is a collective concern for all individuals within the organization and not the sole domain of any particular individual

Table 3. Highest code each category.

Category	Code	Cases/15	% Cases
Security risks organization are facing	User lack of Security Awareness	14	93.3%
Cyberattacks on individuals	Social Engineering	14	93.3%
Threat to security in organization	Insider Threat	8	53.3%
Problem solving guidelines	Building a security behavior	15	100%

5 Conclusion

The issue of cyber security is not confined solely to technical challenges but extends to personnel related concerns within organizations, which are increasingly targeted by cyber threats due to a lack of awareness. The researchers conducted a qualitative data analysis through expert interviews to explore avenues for strategizing towards enhancing organizational cybersecurity knowledge. This investigation focused on four primary issues, including current organizational risks, cyber threats targeting users, cyber threats against the organization itself, and systematic remediation approaches based on expert recommendations. This study reveals that fostering cyber awareness behaviors is unanimously recognized by experts as pivotal for bolstering organizational cyber security effectiveness.

However, this research has limitations. The interviewees, comprising cybersecurity practitioners from both management and operational departments were selected based on their experience and expertise in the field. Nevertheless, cyber awareness is pertinent to all individuals within an organization not solely those in cybersecurity roles. In the future, the researcher plans to interview a sample group of computer users from various organizational departments such as Human Resource, Financial Accounting, Purchasing, Marketing and others to obtain diverse perspectives and insights, thereby enriching the understanding of this topic.

References

1. Kim, A.: 2022 Verizon DBIR – What Does it Mean? | SANS Institute. https://www.sans.org/blog/2022-verizon-dbir-what-does-it-mean/. Accessed 09 Jan 2024
2. Mansfield-Devine, S.: IBM: cost of a data breach. Netw. Secur. **2022**(8) (2022). https://doi.org/10.12968/s1353-4858(22)70049-9
3. Witsenboer, J.W.A., Sijtsma, K., Scheele, F.: Measuring cyber secure behavior of elementary and high school students in the Netherlands. Comput. Educ. **186**, 104536 (2022). https://doi.org/10.1016/j.compedu.2022.104536
4. Shukla, S.S., Tiwari, M., Lokhande, A.C., Tiwari, T., Singh, R., Beri, A.: A comparative study of cyber security awareness, competence and behavior. In: 2022 5th International Conference on Contemporary Computing and Informatics (IC3I), India (2022)

5. Perera, S., Jin, X., Maurushat, A., Opoku, D.-G.J.: Factors affecting reputational damage to organizations due to cyberattacks. Informatics **9**(1), 28 (2022). https://doi.org/10.3390/informatics9010028
6. Wilson, M.G., De Zafra, D.E., Pitcher, S.I., Tressler, J.D., Ippolito, J.B.: Information technology security training requirements. https://doi.org/10.6028/nist.sp.800-16. Accessed 19 Feb 2024
7. Aloul, F.A.: The need for effective information security awareness. J. Adv. Inf. Technol. **3** (2012). https://doi.org/10.4304/jait.3.3.176-183
8. Maraj, A., Butler, W.: Taxonomy of social engineering attacks: a survey of trends and future directions. In: The 17th International Conference on Cyber Warfare and Security, vol. 17, pp. 185–193 (2022). https://doi.org/10.34190/iccws.17.1.40
9. Daengsi, T., Pornpongtechavanich, P., Wuttidittachotti, P.: Cybersecurity awareness enhancement: a study of the effects of age and gender of Thai employees associated with phishing attacks. Educ. Inf. Technol. **27**, 4729–4752 (2021). https://bit.ly/48o2bnj
10. Fujs, D., Mihelič, A., Vrhovec, S.L.R.: The power of interpretation. In: Proceedings of the 14th International Conference on Availability, Reliability and Security (2019). https://bit.ly/42L57ci
11. Alshaikh, M., Maynard, S.B., Ahmad, A., Chang, S.: An exploratory study of current information security training and awareness practices in organizations. In: Proceedings of the 51st Hawaii International Conference on System Sciences (2018). https://bit.ly/3UNovng
12. Gale, N.K., Heath, G., Cameron, E., Rashid, S., Redwood, S.: Using the framework method for the analysis of qualitative data in multi-disciplinary health research. BMC Med. Res. Methodol. **13** (2013). https://doi.org/10.1186/1471-2288-13-117
13. Erlingsson, C., Brysiewicz, P.: A hands-on guide to doing content analysis. Afr. J. Emerg. Med. **7**, 93–99 (2017). https://doi.org/10.1016/j.afjem.2017.08.001
14. Chowdhury, N., Katsikas, S., Gkioulos, V.: Modeling effective cybersecurity training frameworks: a delphi method-based study. Comput. Secur. (2022). https://doi.org/10.1016/j.cose.2021.102551
15. DoD Approved 8570 Baseline Certifications. https://bit.ly/3wwWMgc. Accessed 19 Feb 2024
16. Knapp, K.J., Franklin Morris, R., Marshall, T.E., Byrd, T.A.: Information security policy: an organizational-level process model. Comput. Secur. **28**, 493–508 (2009). https://doi.org/10.1016/j.cose.2009.07.001
17. Timonen, V., Foley, G., Conlon, C.: Challenges when using grounded theory. Int. J. Qual. Methods **17** (2018). https://doi.org/10.1177/1609406918758086
18. Agustianingsih, R., Mahmudi, A.: How to design open-ended questions?: literature review. In: Journal of Physics: Conference Series (2019)
19. Yerby, J., Floyd, K.: Faculty and staff information security awareness and behaviors. J. Colloquium Inf. Syst. Secur. Educ. (CISSE) **6**, 138–160 (2018)
20. Almrezeqa, N., Alserhani, F., Humayun, M.: Exploratory study to measure awareness of cybercrime in Saudi Arabia. Turk. J. Comput. Math. Educ. (TURCOMAT) (2021)
21. Grassegger, T., Nedbal, D.: The role of employees' information security awareness on the intention to resist social engineering. Procedia Comput. Sci. **181**, 59–66 (2021). https://doi.org/10.1016/j.procs.2021.01.103
22. Chen, Y., Ramamurthy, K., Wen, K.-W.: Impacts of comprehensive information security programs on information security culture. J. Comput. Inf. Syst. **55**, 11–19 (2015). https://doi.org/10.1080/08874417.2015.11645767
23. Byrne, D.: A worked example of Braun and Clarke's approach to reflexive thematic analysis. Qual. Quant. **56**, 1391–1412 (2021). https://doi.org/10.1007/s11135-021-01182-y
24. Almansoori, A., Al-Emran, M., Shaalan, K.: Exploring the frontiers of cybersecurity behavior: a systematic review of studies and theories. Appl. Sci. **13**, 5700 (2023). https://doi.org/10.3390/app13095700

25. Alnsour, M.A.: Using modified grounded theory for conducting systematic research study on sustainable project management field. MethodsX (2022). https://doi.org/10.1016/j.mex.2022.101897
26. Ralph, N., Birks, M., Chapman, Y.: The methodological dynamism of grounded theory. Int. J. Qual. Methods **14**, 160940691561157 (2015). https://doi.org/10.1177/1609406915611576

Human-Motion Guided Frame Selection with Adaptive Gamma Correction for Violent Video Classification

Sorn Sooksatra and Sitapa Watcharapinchai(✉)

National Electronic and Computer Technology Center, National Science and Technology Development Agency, Pathum Thani, Thailand
{sorn.soo,sitapa.wat}@nectec.or.th

Abstract. This paper proposes a human-motion guided frame selection approach for violent video classification. The human-motion features are computed by determining the frame difference within the detected human regions. Additionally, adaptive gamma correction is introduced in the motion signal to mitigate abrupt motion. The experiment is evaluated on RWF-2000 dataset by using I3D network as a classification model. The empirical results demonstrate that the proposed method outperforms the state-of-the-art methods in violent classification. Selecting more informative frames can improve the classification performance compared to traditional frame selection methods that use uniform sampling. Therefore, the proposed method enables the extraction of more informative frames.

Keywords: Frame Selection · Violent Video Classification · Gamma Correction · Human Detection

1 Introduction

In recent years, violent video classification has gained attention in action classification. This method has a role to classify violent scenes from trimmed surveillance videos. Typically, classification accuracy can be improved by focusing on the source of action or human regions [1]. Despite significant progress in violent video classification, frame selection can still be further improved to extract more informative frames. In general, frame selection in violent video classification utilizes a uniform sampling [2], as a baseline technique, where frames are chosen uniformly along time dimension. However, this technique may capture uninformative frames that lack motion information, resulting in poor classification performance, particularly for actions with short duration. Additionally, an abrupt motion caused by camera vibration or a shot change can suppress the salience of the action motion. Therefore, it is essential to be considered for the abrupt motion suppression when performing frame selection.

This paper focuses on enhancing frame selection to improve violent video classification. Considering the characteristics of violent action, most actions are done

P. Meesad et al. (Eds.): IC2IT 2024, LNNS 973, pp. 12–22, 2024.
https://doi.org/10.1007/978-3-031-58561-6_2

by people. Therefore, the proposed method aims to extract informative frames based on human motion in each frame by utilizing frame difference and human detection as human-motion features. To reduce the effect of abrupt motion, the proposed method incorporates adaptive gamma correction [3]. Figure 1 shows three examples that compare frame selection methods using uniform sampling (1st row) and the proposed method (2nd row). It demonstrates the capability of the proposed method to focus on the human motion within bounding boxes as shown in Fig. 1(a). In Fig. 1(b), more background frames without motion can be discarded as denoted by "uninformative frame" in black lines. In addition, adaptive gamma correction helps to reduce selected frames affected by abrupt motion (shot change) as shown in Fig. 1(c). To summarize, our contributions are listed as follows:

- Human-motion guided frame selection was proposed to extract frames focusing on human action or violent behavior.
- Adaptive gamma correction was adapted in motion information to discard frames with the abrupt motion.

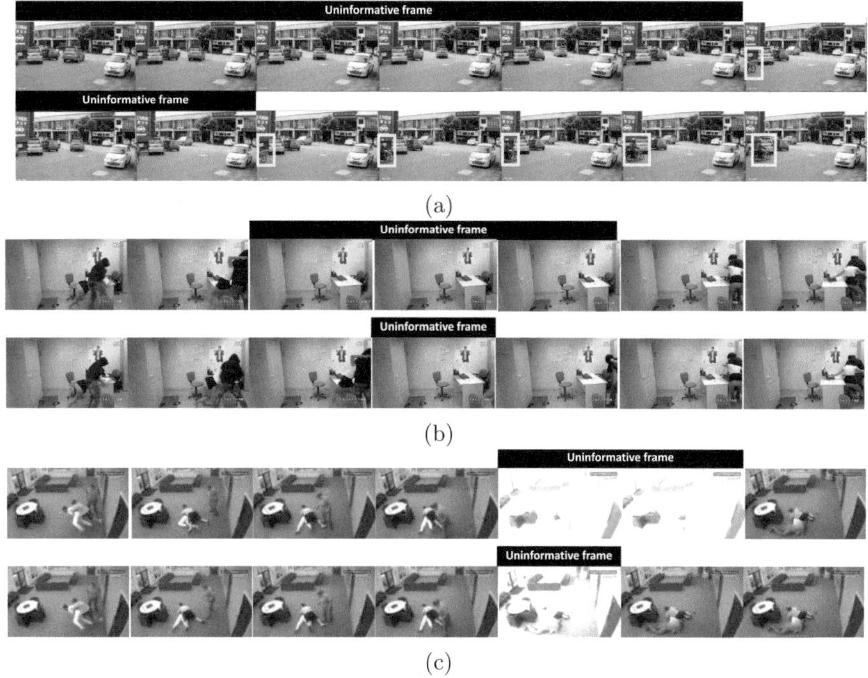

Fig. 1. The examples of frame selection methods using uniform sampling (1st row) and the proposed method (2nd row) in case of (a) focusing on human motion (within bounding boxes), (b) ignorance for non-motion frames, and (c) abrupt motion, where uninformative frames are highlighted with black color.

2 Related Work

For the past decade, there have been several studies related to action classification, with a particular emphasis on violent behavior and the design of models to classify violent videos. Frame selection has also been explored to identify informative frames for analysis. This section provides brief contributions from related studies in violent video classification and frame selection.

2.1 Violent Video Classification

Most recent violent video classification relied on deep-learning-based techniques. Several cues were utilized for detecting violent scenes. Two-stream network [4,5] from RGB frames and motion cues was applied for classification. Since violent action was related to the human, skeleton data from the SPIL module [6] was proposed to capture both features and positions relation information simultaneously. Self-attention mechanism from vision transformer [7] was able to extract skeleton features and increase classification performance. Besides motion and human cues, U-shape architecture [8] could be applied to reuse low-level features.

2.2 Frame Selection for Action Classification

Since video sequences usually contained lots of redundant frames, frame selection played an important role in selecting informative frames for reducing computation cost. Frame selection was first attempted in TSN [2] through frame sampling with equal duration or uniform sampling. However, selected frames might be uninformative or non-motion frames. Recently, deep-learning techniques were utilized in frame selection. Reinforcement learning [9] was applied to select key and informative frames in untrimmed videos with long duration. LSTM [10] gathered a global memory with a duplicated frame over time trained by gradient methods. SCSampler [11] relied on a lightweight CNN to generate salience scores for frame selection. A hierarchical search algorithm with feature mapping [12] effectively selects frames based on the search and learning algorithm. A dynamic version of TSN [13] was proposed with Reinforcement learning. To reduce computation cost, handcrafted techniques were proposed in this research. Multiple frames could be aggregated into single frames by Gaussian Weighting Function [14]. Motion cues generated by a simple frame difference were used for frame selection in MG-sampler [15].

Compared to deep-learning-based techniques in frame selection, the proposed method relied on a simple handcrafted technique for trimmed violent videos. Even though MG-sampler showed sufficient performance without deep-learning-based techniques, further improvements for violent video classification can be developed regarding abrupt and human motion, described in the next section.

3 Overall Proposed Architecture

The proposed method aims to extract informative frames for violent video classification, using RGB frames. Let $\{I[0], I[1], ..., I[T]\}$ be the set of RGB frames as the input data for frame selection method, where T was the total number of frame in videos. The overall architecture was illustrated in Fig. 2 consisting of two parts (Human-motion features and Human-motion guided frame selection). The first part was utilized to extract motion features focusing on people through frame difference. The human region of interest (ROI_h) was obtained from human detection process. Then, the features were generated into cumulative motion distribution. The second part obtained and analyzed the motion distribution for selecting informative frames. Finally, all selected frames are fed into a classification model for violent classification.

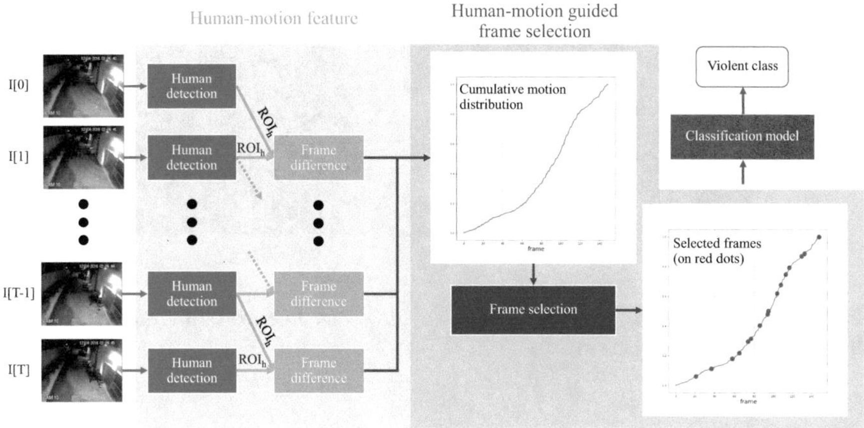

Fig. 2. The illustration of overall model architecture consisting of human-motion feature and human-motion guided frame selection.

4 Human-Motion Feature

In this section, we focused on generating human-motion features in each frame. Although optical flows [16] was widely used as motion features over the past decade, they often came with a high computation cost for video recognition. Based on MG-sampler, the proposed method utilizes Structural Similarity [17] to compute frame difference as motion features, as shown in Eq. (1)

$$D[i] = 1 - ssim(I[i], I[i-1]) \tag{1}$$

where $I[i]$ was the ith RGB frame from the input video. $D[i]$ was the ith frame difference from Structural Similarity ($ssim$), generating scores from 0 (different) to 1 (similar). $D[0]$ was set as 0 in the experiment. On the other hand, YOLO [18] was utilized as human detection to extract human regions shown in Fig. 2. The human motion could be obtained within these regions. Based on our observation, people tend to not move far in short-duration scenes or trimmed videos (<5 s) and tend to stay in the near positions throughout the time as shown in Fig. 3(a). Additionally, human detection may be missed in some frames. Thus, we decided to merge all human detected regions (Fig. 3(b)) in the entire sequence of the input video as ROI_h as depicted in Fig. 3(c). Therefore, the input images ($I_h[i]$) for human motion were calculated by multiplication between $I[i]$ and ROI_h. Equation (2) shows a calculation for the human-motion feature as follows:

$$D_h[i] = 1 - ssim(I_h[i], I_h[i-1]) \tag{2}$$

where $D_h[i]$ was the ith frame difference for human-motion features and $D_h[0]$ was set as 0 in this experiment.

(a) (b) (c)

Fig. 3. The example of (a) input image, (b) detected human bounding boxes in the whole sequence, and (c) its ROI_h.

5 Human-Motion Guided Frame Selection

After obtaining the human-motion features from each frame, their scores were utilized as frame selection criteria described in this section. Inspired by MG-sampler [15], informative frames could be selected from the slope of cumulative motion distribution representing high values of frame difference calculated by Eq. (3)

$$C[t] = \frac{\sum_{i=0}^{t} D^{\gamma}[i]}{\sum_{i=0}^{T-1} D^{\gamma}[i]} \tag{3}$$

where $C[t]$ was the cumulative motion distribution at the tth frame and γ represented as a smoothness parameter. T was the total number of frames in the input video. In the previous work [15], a fixed value of γ was applied to control the smoothness of the cumulative motion distribution. With a lower value of γ, the frames were selected more uniformly. In this paper, adaptive gamma correction [3] was modified to find a suitable value of γ and remove an abrupt peak or motion. The concept of this technique was used in a signal enhancement so that the average value of the transformed signal is $\frac{1}{2}$. Therefore, the adaptive γ was calculated by Eq. (4)

$$\gamma = \frac{\ln(1/2)}{\ln(D_{avg})} \tag{4}$$

where D_{avg} was an average value of D. Figure 4 shows examples of signals with (left) and without abrupt peaks (right). It indicated that an abrupt signal was suppressed in a transformed signal by adaptive gamma correction. On the other hand, this technique still kept an original form for a signal without abrupt peaks.

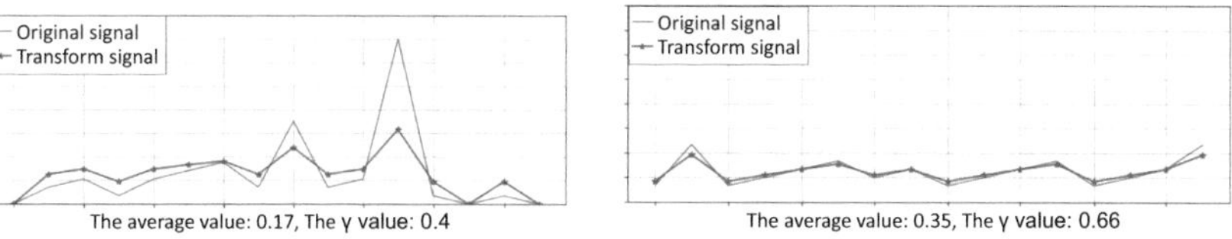

Fig. 4. Examples of original and transformed signals using adaptive gamma correction, with (left) and without abrupt peaks (right).

In order to select N frames from C, the criteria were followed from MG-sampler [15] by dividing N temporal parts with N thresholds. Let F be the set of selected frames $\{F[1], F[2], ..., F[n], ..., F[N]\}$ and the nth threshold represented as $\frac{1+2n}{2N}$. Each frame was selected in the value of C which was the closest to each threshold as shown in Eq. (5)

$$F[n] = \underset{t}{\text{argmin}}\left(abs(C[t] - \frac{1+2n}{2N})\right) \tag{5}$$

6 Experiments

6.1 Experimental Setup

Dataset: The videos in this experiment were obtained from RWF-2000 [4]. This dataset contains 2000 videos which are violent and non-violent scenes. They are trimmed from surveillance-only videos, each with a duration of 5 s and a frame rate of 30 fps. Videos were divided into training and validation sets consisting of 1600 and 400 videos, respectively. The classification model was evaluated on the validation set. The number of violent and non-violent scenes in this experiment were equally divided for each set.

Implementation Detail: In this experiment, the I3D network (RGB only) [19] was employed as the classification model to calculate classification accuracy for frame selection. This model also was utilized for analyzing the behavior of frame selection by fine-tuning from the training data of RWF-2000 dataset in each snippet. The I3D network was optimized using Adam optimizer with learning rate = 10^{-4} and multiplicative factor of learning rate decay = 0.1. The cross-entropy error was utilized as the loss function.

Evaluation Metric: The criteria was followed from most related works in frame selection. Indirect evaluation was performed by assessing the action classification accuracy [1]. In addition, frame selection behavior was analyzed with violent probabilities for quality assessment. Frame selection behavior was analyzed with violent probability generated from fine-tuned classification model.

6.2 Ablation Study

This section described the ablation study conducted to evaluate frame selection. We utilized uniform sampling ($\gamma = 0$) and the frame difference (F_D) MG-sampler ($\gamma = 1$) [15] as the baseline technique in this experiment. The proposed method had two configurations, frame difference with human-motion features ($F_{D,H}$) and adaptive gamma correction (γ_{Agc}). Table 1 showed classification accuracy, ranging the number of selected frame (N) from 16 to 64 frames. It clearly showed that MG-sampler outperformed uniform sampling by 2%. In low number of N (16 to 32), human-motion features ($F_{D,H}$) outperformed other configurations. Input features with adaptive gamma correction yielded the best accuracy at a higher N (starting from 48 frames). It showed that abrupt motion suppression was effective for violent video classification. In addition, it had a significant improvement of 2.75% compared with using a fixed γ value. However, the accuracy from the combination between $F_{D,H}$ and γ_{Agc} was slightly decreased. The decline in accuracy from this combination might be caused by sudden human motion which was suppressed by adaptive gamma correction. Informative frame, especially in frames with human motion, might be ignored for frame selection in such cases. Therefore, the result indicated that $F_{D,H}$ with γ_{Agc} was less effective for this combination.

Table 1. The classification accuracy for frame selection with various configurations.

Input feature	γ value	Classification accuracy (%)			
		N = 16	N = 32	N = 48	N = 64
Uniform sampling	0	80.60	85.00	85.75	85.75
MG-sampler	1	83.50	85.00	86.25	87.75
$F_{D,H}$	1	**84.75**	**88.50**	**89.25**	90.00
F_D	γ_{Agc}	**84.75**	88.00	**89.25**	**90.50**
$F_{D,H}$	γ_{Agc}	83.50	86.75	88.00	89.25

6.3 Comparative Result on Violent Video Classification

The experiment results were obtained from the state-of-the-art (SOTA) related to violent video classification with uniform sampling. The validation set of RWF-2000 was used for evaluation. The classification accuracy was obtained from their publications. Table 2 showed the comparative result between the SOTAs and the proposed method with three configurations (F_D with γ_{Agc}, $F_{D,H}$ with $\gamma = 1$, and $F_{D,H}$ with γ_{Agc}). The number selected frame was set to 64. Table 2 showed that the proposed method outperformed other SOTAs in violent video classification around 1%, compared to F_D with γ_{Agc}. This study has shown that frame selection could be an important factor in enhancing violent video classification performance.

Table 2. Comparative results to the state-of-the-art on a validation set of RWF-2000.

Model name	Accuracy (%)
SepConvLSTM [5]	89.75
SPIL Convolution [6]	89.30
TokenPose-B+3D-CNN [7]	89.45
Efficient network [8]	85.00
Flow Gated Network [4]	87.25
Proposed method ($F_{D,H}$ with $\gamma = 1$)	90.00
Proposed method (F_D with γ_{Agc})	**90.50**
Proposed method ($F_{D,H}$ with γ_{Agc})	89.25

6.4 Analysis on Behavior of Frame Selection

This section analyzed the behavior of frame selection in various configurations. We compared selected frames (blue lines) and their violent probabilities in a vertical axis (red lines) for each temporal position in a horizontal axis as shown in Fig. 5, depicting three video samples in each row. In our hypothesis, informative frames should be located high violent probabilities. Figure 5 clearly showed that frames selected by most methods, especially in MG-sampler and $F_{D,H}$ with $\gamma = 1$ were located at high violent probabilities. As can be seen in the 3rd column of Fig. 5, $F_{D,H}$ was able to draw temporal boundaries between informative and uninformative frames. Even though frame selection with γ_{Agc} outperformed other configurations as shown in Table 1, it was also able to spread selected frames for the entire video with the focus on informative frames as illustrated in the 4th and 5th columns of Fig. 5. It is also important to include some uninformative frames to provide context and enrich violent features.

The third row of Fig. 5 depicted a violent scene with equal violent probabilities. In this case, it also had abrupt motion during 0.5 and 0.7 in the 2nd column of Fig. 5. The result from $F_{D,H}$ indicated that the human motion might be less salient than abrupt motion. Then, their selected frames could not extract all informative frame. For this circumstance, the proposed method with γ_{Agc} could be considered to remove abrupt motion and focused on informative frames.

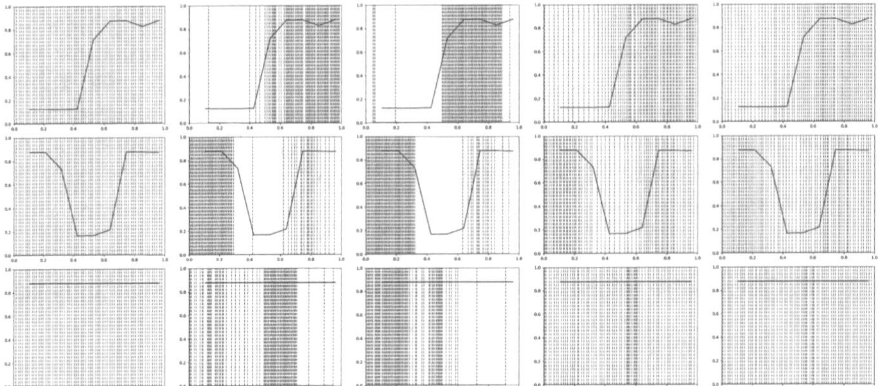

Fig. 5. The examples of violent probabilities (red lines) with 64 selected frame (blue dotted lines) from three violent videos in each row, utilizing uniform sampling (1st column), MG-sampler (2nd column), $F_{D,H}$ with $\gamma = 1$ (3rd column), F_D with γ_{Agc} (4th column), and $F_{D,H}$ with γ_{Agc} (5th column), where vertical and horizontal axes were violent probabilities and temporal position, respectively.

7 Conclusion

This paper proposed a frame selection for violent video classification. In the proposed method, informative frames could be extracted by focusing on human-motion features and suppressing abrupt motion by adaptive gamma correction. Empirical results revealed that the proposed method outperformed the baseline techniques (uniform sampling and MG-sampler) and other SOTAs approximately 3% and 1%, respectively. On the other hand, the selected frames from our proposed method were able to identify the boundaries between violent and non-violent scenes in quality assessment, especially with human motion, indicating that better informative frames were effectively selected. In the future work, further improvements could be made to combine human-motion features with adaptive gamma correction more efficiently. Additionally, incorporating object motion by leveraging human-to-object interaction techniques could enhance the frame selection process.

Acknowledgments. This research has received funding support from the NSRF via the Program Management Unit for Human Resources & Institutional Development, Research and Innovation (grant number B04G640053). Computational resources for this work were supported by NSTDA Supercomputer Center (ThaiSC).

References

1. Mumtaz, N., et al.: An overview of violence detection techniques: current challenges and future directions. Artif. Intell. Rev. **56**, 4641–4666 (2023). https://doi.org/10.1007/s10462-022-10285-3
2. Wang, L., et al.: Temporal segment networks for action recognition in videos. IEEE Trans. Pattern Anal. Mach. Intell. **41**(11), 2740–2755 (2018)
3. Yelmanov, S., Romanyshyn, Y.: A new technique for image enhancement by adaptive power-law intensity transformation. In: 2021 IEEE 12th International Conference on Electronics and Information Technologies (ELIT), pp. 90–94. IEEE (2021)
4. Cheng, M., Cai, K., Li, M.: RWF-2000: an open large scale video database for violence detection. In: 2020 25th International Conference on Pattern Recognition (ICPR), pp. 4183–4190 (2021). https://doi.org/10.1109/ICPR48806.2021.9412502
5. Islam, Z., Rukonuzzaman, M., Ahmed, R., Kabir, M.H., Farazi, M.: Efficient two-stream network for violence detection using separable convolutional LSTM. In: 2021 International Joint Conference on Neural Networks (IJCNN), pp. 1–8. IEEE (2021)
6. Su, Y., Lin, G., Zhu, J., Wu, Q.: Human interaction learning on 3D skeleton point clouds for video violence recognition. In: Vedaldi, A., Bischof, H., Brox, T., Frahm, J.-M. (eds.) ECCV 2020. LNCS, vol. 12349, pp. 74–90. Springer, Cham (2020). https://doi.org/10.1007/978-3-030-58548-8_5
7. Zhou, L.: End-to-end video violence detection with transformer. In: 2022 5th International Conference on Pattern Recognition and Artificial Intelligence (PRAI), pp. 880–884. IEEE (2022)
8. Vijeikis, R., Raudonis, V., Dervinis, G.: Efficient violence detection in surveillance. Sensors **22**(6), 2216 (2022)
9. Fan, H., Xu, Z., Zhu, L., Yan, C., Ge, J., Yang, Y.: Watching a small portion could be as good as watching all: towards efficient video classification. In: IJCAI International Joint Conference on Artificial Intelligence (2018)
10. Wu, Z., Xiong, C., Ma, C.Y., Socher, R., Davis, L.S.: AdaFrame: adaptive frame selection for fast video recognition. In: Proceedings of the IEEE/CVF Conference on Computer Vision and Pattern Recognition, pp. 1278–1287 (2019)
11. Korbar, B., Tran, D., Torresani, L.: SCSampler: sampling salient clips from video for efficient action recognition. In: Proceedings of the IEEE/CVF International Conference on Computer Vision, pp. 6232–6242 (2019)
12. Zhao, M., Yu, Y., Wang, X., Yang, L., Niu, D.: Search-map-search: a frame selection paradigm for action recognition. In: Proceedings of the IEEE/CVF Conference on Computer Vision and Pattern Recognition (CVPR), pp. 10,627–10,636 (2023)
13. Zheng, Y.D., Liu, Z., Lu, T., Wang, L.: Dynamic sampling networks for efficient action recognition in videos. IEEE Trans. Image Process. **29**, 7970–7983 (2020)
14. Basha, S.S., Pulabaigari, V., Mukherjee, S.: An information-rich sampling technique over spatio-temporal CNN for classification of human actions in videos. Multimed. Tools Appl. **81**(28), 40431–40449 (2022)
15. Zhi, Y., Tong, Z., Wang, L., Wu, G.: MGSampler: an explainable sampling strategy for video action recognition. In: Proceedings of the IEEE/CVF International Conference on Computer Vision, pp. 1513–1522 (2021)
16. Teed, Z., Deng, J.: RAFT: recurrent all-pairs field transforms for optical flow. In: Vedaldi, A., Bischof, H., Brox, T., Frahm, J.-M. (eds.) ECCV 2020. LNCS, vol. 12347, pp. 402–419. Springer, Cham (2020). https://doi.org/10.1007/978-3-030-58536-5_24

17. Wang, Z., Bovik, A.C., Sheikh, H.R., Simoncelli, E.P.: Image quality assessment: from error visibility to structural similarity. IEEE Trans. Image Process. **13**(4), 600–612 (2004)
18. Wang, C.Y., Bochkovskiy, A., Liao, H.Y.M.: YOLOv7: trainable bag-of-freebies sets new state-of-the-art for real-time object detectors. arXiv preprint arXiv:2207.02696 (2022)
19. Carreira, J., Zisserman, A.: Quo vadis, action recognition? A new model and the kinetics dataset. In: Proceedings of the IEEE Conference on Computer Vision and Pattern Recognition, pp. 6299–6308 (2017)

Predictive Modeling of Vicat Softening Point for Low-Density Polyethylene Using GBM, XGBoost and AdaBoost: A Comparative Analysis

Noparat Phongthakun, Sunisa Rimcharoen, and Nutthanon Leelathakul(✉)

Faculty of Informatics, Burapha University, Chon Buri 20131, Thailand
{65810010,rsunisa,nutthanon}@go.buu.ac.th

Abstract. This study provides a comparative analysis for predicting Vicat softening temperatures of Low-Density Polyethylene (LDPE), which is one of the versatile polymers used extensively across various industrial sectors. LDPE exhibits unique properties, such as flexibility, electrical insulating characteristics, and low melting point. The manufacturing process demands rigorous quality controls, involving extensive laboratory product testing, requiring significant time, labor, and cost investments. In this study, we explored the potential of machine-learning-based predictors to alleviate or reduce these challenges. Our analysis focused on the accuracy and processing time of predicting models based on three prominent boosting methods: Gradient Boosting Machines (GBM), Extreme Gradient Boosting (XGBoost), and Adaptive Boosting (AdaBoost). We collected the laboratory testing results from one of the largest polymer manufacturers in Southeast Asia: our data set comprised 71 features. Based on the comparison results, we concluded that XGBoost exhibits superior predictive performance (in terms of MAE, MSE, and RMSE) compared to GBM and AdaBoost, indicating its potential in time saving, labor, and cost reduction in the manufacturing process. Both XGBoost and AdaBoost incurred maximum errors below 2.9, aligning with the industry testing standard. Notably, Adaboost incurred slightly lower maximum errors in comparison to XGBoost. Furthermore, we presented the top 10 significant features highlighted by the XGBoost models.

Keywords: GBM · XGBoost · AdaBoost · Prediction · Low-Density Polyethylene · Quality Control

1 Introduction

Low-Density Polyethylene (LDPE) has played a significant role across various industries due to its unique properties. LDPE's flexibility, electrical insulating characteristics, chemical resistance, and low melting point has made it a preferred material for diverse applications, such as packaging materials, medical devices, agricultural films, and consumer products [1].

P. Meesad et al. (Eds.): IC2IT 2024, LNNS 973, pp. 23–32, 2024.
https://doi.org/10.1007/978-3-031-58561-6_3

The significance of predicting LDPE parameters lies in three key aspects. Firstly, the prediction can reduce reliance on time-consuming and labor-intensive laboratory tests [2]. Secondly, it could facilitate the fine-tuning of LDPE formulations, leading to enhanced product performance and cost effectiveness. Lastly, it supports sustainability efforts by facilitating the LDPE production processes with reduced carbon footprints and lower energy consumption.

Vicat softening temperature is a crucial parameter [3] in the selection of materials for specific applications. Knowing this temperature allows engineers and designers to make informed decisions about the suitability of a material for a particular environment or use, ensuring the safety and performance of products. This ensures that the materials used will withstand the required conditions and remain stable over time, reducing the likelihood of product failure or premature deterioration. Manufacturers can also use this parameter to monitor and maintain consistent material properties in their production processes, reducing defects and ensuring product reliability.

However, the Vicat softening temperature is determined through a standardized testing procedure, typically following the guidelines outlined in international standards such as ASTM D1525 or ISO 306. The process includes sample preparation, testing, and data analysis, which can typically take a few hours to a few days per sample. Costs can vary widely depending on various factors such as government subsidy, material types, testing conditions, and laboratory quality controls.

This paper, therefore, comparatively analyzes machine learning algorithms for predicting Vicat softening temperatures of LDPE products, aiming to provide an alternative to traditional laboratory-based testing. The selected algorithms learn to estimate Vicat softening temperatures by examining real-world data from a factory comprising diverse properties and characteristics inherent to LDPE products. The experimental results show that the model derived from the study holds the promise of substantial time, labor, and cost savings in the determination of Vicat softening temperatures.

The contributions of this research, focusing on predicting the Vicat softening temperature by leveraging GBM, XGBoost, and AdaBoost, lie in four key aspects. Firstly, the comparative analysis shows the machine learning-based model that outperforms the others, and is suitable for predicting the values of Vicat softening temperature. Secondly, the best-performing parameters of each algorithm are reported. Next, the top 10 important features, based on the best-performing predictor, are discovered, analyzed, and confirmed by the expert chemists. The value of feature importance is crucial for understanding which features are most influential in the model's predictions and can guide further feature selection and optimization efforts. Lastly, the prediction model is constructed by utilizing real operational data collected over a three-year period from an active industrial facility. This concrete dataset not only validates the viability of our prediction model but also positions it as a dependable solution for real-world applications, demonstrating its readiness for practical deployment.

Section 2 summarizes the related works including various forecasting methods, the effectiveness of boosting-based approaches in different domains and its potential in the polymer industry, and previous methods for predicting the Vicat softening temperature. In Sect. 3, we explain our methodology, covering aspect such as data collection, data analysis, data preparation, and model selection. The experiment results, detailed in Sect. 4, show the most effective hyperparameters and errors for different ensemble learning techniques (GBM, XGBoost, and AdaBoost). Additionally, Sect. 4 highlights the top 10 important features correlated to the values of the Vicat softening temperatures of LDPE.

2 Related Work

We explored various forecasting methods applied to a wide range of domains, including the prediction of electrical consumption, market demand, stock market price, power loads of data centers, temperatures, wheat yield, electricity demand in liberalized markets, and electricity spot price forecasting [4–11]. A variety of techniques, including machine learning, statistical, and ensemble approaches, were proposed. In [4], a time series technique called SARIMA together with arithmetic averages, forming an ensemble, to predict electrical consumption. [5–7] depended on time series-based approaches to predict data-center power load, electricity demand, and electricity price, respectively. Neural networks utilized in [6, 8, 9] were used to predict transformer load, temperature in New Delhi, e-commerce demand, and power system load, respectively. LSTM-based approaches proposed in [10] and [11] in order to forecast demand of dairy products and of fresh-food logistics, respectively. Exponential Smoothing, and K-Nearest Neighbor approaches were proposed for forecasting oil, coal, and natural gas prices in India, both pre- and post-COVID-19 [7].

Boosting-based approaches have been particularly effective in diverse domains because they can handle not only a variety of data types, including categorical and numerical data, but also missing value and outliers. Among boosting-based methods, prior research has indicated that the classical Adaboost and widely-used XGBoost are promising for making predictions. In [12], the study highlighted the effectiveness of the AdaBoost algorithm in reducing overfitting and improving performance, demonstrating its potential in predicting water temperature. In [13], an ensemble of 7 machine learning algorithms (such as XGBoost, random forest, and support vector regression) was used to predict wheat yields. Moreover, previous studies showed that the XGBoost algorithm was particularly effective in diverse domains, including estimating surface water quality [14], forecasting retail sales [15], wave height forecasting [16], developing predictive maintenance models for wind turbines [17], interpreting factors affecting crude oil prices [18], and predicting software maintenance efforts [19]. Nevertheless, XGBoost has not yet been employed for predicting characteristics of LDPE using real-world data obtained from manufacturing facilities.

The most related to our proposed model, [2] proposed predictive models, including Linear regression, Support Vector Regression (SVR), and Adaboost Regressor, for forecasting the melt flow index (MFI), in contrast to our focus on predicting the Vicat softening temperature. Predicting the Vicat softening temperature is vital as the laboratory product testing process requires more manhours and is time consuming compared to the MFI testing. Consequently, accurate Vicat predictions offer the potential for more cost reduction.

3 Methodology

Figure 1 illustrates the process of our methodology, starting from data collection, followed by data preparation, model selection, and concluding with model evaluation.

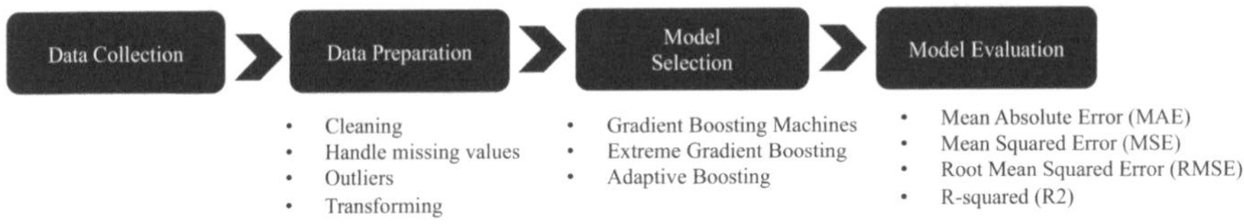

Fig. 1. Overview of our research methodology.

3.1 Data Collection and Analysis

Figure 2 shows the system diagram of the LAB Information Management System (LIMS). After a lab analyst conducts a test on an LDPE Product, the testing instrument transmits the test results to a specific lab computer. Subsequently, the data is transferred to LIMS. Additionally, the test results can be accessed or exported in the CSV format, through the Reporting Server.

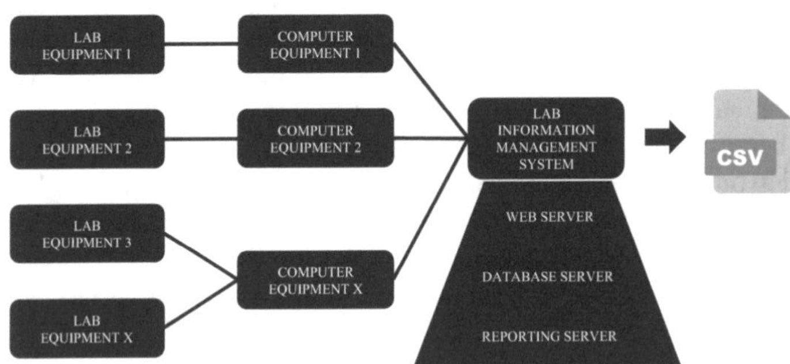

Fig. 2. Lab information management system (LIMS).

This research study utilized real data from an LDPE manufacturing plant, which comprised approximately 4500 rows. The data were collected during a period of 5 years (2018–2023). The data set contained 71 features (65 of which are numerical and 6 of which are categorical). The number of rows that had the testing values of Vicat was 181. Table 1 shows the example of the features together with their type, used in our experiments.

Table 1. Example of the characteristics of LDPE product.

Features	Type	Features	Type
Density	Numerical	Tensile at break	Numerical
Tensile at Break (Machine Direction)	Numerical	Product grade	Categorical
Gloss 20o	Numerical	Blown film	Categorical
MFR 190/2.16 (ISO)	Numerical	Other defects	Categorical
COF-Kinetic(48 h)	Numerical	Foreign details	Categorical

3.2 Data Preparation

To prepare data, a process of data cleansing was conducted, along with other processes. These processes were necessary because the data had been accumulated over several years and had undergone multiple system migrations, causing various formats and missing values.

The initial dataset contained 4500 rows and 71 columns, of which example rows and features are shown in Fig. 3. Data preparation involved several steps to clean and pre-process the data. The first step involved deleting the rows without the Vicat values: only 181 rows remain. The second step involved filling any missing values of the other numerical features with their mean, and of the categorical features with the most frequent category. Subsequently, the third step involved the deletion of duplicate 10 columns, resulting in the D1 dataset (consisting of 181 rows and 61 columns). Finally, we removed the columns whose more than 70% of values are zeros, yielding the D2 dataset (comprising 181 rows and 41 columns).

Example Features	Description	Units	Example Data Points					
			1	2	3	4	5	6
%ASH	Percentage of ash content	% of weight	0.08		0.07			0.1
2 < PSD < 4.0 mm	Particle size distribution	%		99.85	99.99			99.9
ASH	Amount of ash content	ppm	771.5		738.6			1034
Blown film	Completed if the color of the final product is clear (not brown)	-	Com pleted	Com pleted	Com pleted	Com pleted		
COF-Kinetic(48hrs)	Coefficient of friction - Kinetic after 48 hours	-	0.09		0.1			0.12
COF-Static(48hrs)	Coefficient of friction - Static after 48 hours	-	0.1		0.13			0.14
Defect detail	Yes if the final product is defective	-	No	No	No	No	Yes	No
Density g/cm3	Density in grams per cubic centimete	g/cm3	0.92	0.92	0.92	0.92	0.92	0.93

Fig. 3. Examples of the features in the data set.

3.3 Model Selection

Based on both XGBoost and AdaBoost, the comparative studies in this research aimed to harness their predictive power to predict LDPE parameters. XGBoost and AdaBoost have emerged as powerful tools for precise prediction and modeling of material properties. AdaBoost focuses on boosting the performance of weak predictors by adjusting their weights to form a stronger predictor. It has been effective in reducing overfitting and improving predictive performance across various domains. XGBoost, known for its adaptability and resilience in handling complex data, has gained recognition for its exceptional predictive capabilities in various fields. It utilizes an ensemble learning approach, combining predictions from multiple decision tree models to deliver robust and highly accurate results. Its versatility in handling regression and classification tasks, coupled with its ability to manage intricate and high-dimensional data without overfitting, makes XGBoost an ideal choice for predicting LDPE parameters.

Therefore, we opted to employ XGBoost and AdaBoost for predicting the Vicat softening temperature, and conducted a comparative analysis with the performance of the traditional boosting method, GBM.

3.4 Model Evaluation

Evaluating the model is a crucial step in assessing the performance of the selected algorithms. Various metrics, including Mean Absolute Error (MAE), Mean Squared Error (MSE), Root Mean Squared Error (RMSE), R-squared (R^2) as in [18], along with maximum value of error and processing time, were considered. Cross-validation was also employed to provide a more reliable performance estimate, reducing the risk of overfitting and ensuring the model's robustness in real-world applications.

4 Experimental Results

The specifications of the machine used in these experiments were 3.1 GHz 6-Core Intel Core i5, Radeon Pro 575X 4 GB, and 32 GB 2667 MHz DDR4. Table 2 provides the ranges of hyperparameters that were set for GBM, XGBoost, and AdaBoost. The values marked with asterisks (*) were the best performing ones, and "N/As" represented the hyperparameters not used by the methods. For 'n_estimators', which denotes the number of boosting stages or trees in the ensemble, the best performing values were 100 for GBM, 50 for XGBoost, and 100 for AdaBoost. Regarding the 'max_depth', representing the maximum depth of individual trees, the best performing values were 5 for GBM and 7 for XGBoost. For 'min_child_weight', which determines the child's minimum sum of instance weights, XGBoost was most effective at the value of 1.1. The 'learning_rate' hyperparameter, influencing the contribution of each tree to the model, yielded most accuracy at 0.55 for GBM, 0.51 for XGBoost, and 0.55 for AdaBoost. Lastly, for the 'loss' function, which impacts how misclassifications are penalized, AdaBoost performed best when 'loss' was set to 'exponential'.

Figure 4 provides a comprehensive evaluation of GBM, XGBoost, and AdaBoost, in terms of Mean Absolute Error (MAE), Mean Squared Error (MSE), Root Mean

Table 2. Summary of hyperparameter.

Hyperparameter	GBM	XGBoost	AdaBoost
n_estimators	50, **100***, 150, 200, 250	**50***, 100, 150, 200, 250	50, **100***, 150, 200, 250
max_depth	4, **5***, 6, 7	4, 5, 6, **7***	N/A
min_child_weight	N/A	**1.1***, 1.2, 1.3, 1.4, 1.5	N/A
learning_rate	0.51, 0.52, 0.53, 0.54, **0.55***	**0.51***, 0.52, 0.53, 0.54, 0.55	0.51, 0.52, 0.53, 0.54, **0.55***
loss	N/A	N/A	linear, square, exponential

Squared Error (RMSE), R-squared (R^2), max errors, and % max error, along with their respective computational time. Each green-highlighted value was the performance of the best-performing method according to one of the metrics. (For example, the XGBoost outperforms the others in terms of MAE as its value was 0.26.) Each underlined value belonged to the best-performing method when the number of features was set to 41, 61, and 71, respectively. (For instance, when considering a scenario with 41 features, AdaBoost outperformed the others in terms of MAE, with the value of 0.36, whereas GBB demonstrated superior performance in terms of MSE with the value of 0.32.)

The results revealed that the D1 dataset (with 61 selected features) incurred the lowest errors, suggesting that the features, whose values of zeroes are predominant, were significant in predicting the Vicat values. For instance, "Snake skin" was an indisposable feature for the prediction, where its value of 0 indicated whether the product surface was smooth.

Based on the D1 dataset, XGBoost outperformed the others. It slightly had lower errors than AdaBoost: 0.26, 0.11, 0.33, and 0.98, in terms of MAE, MSE, RMSE, R2, respectively, as opposed to those of AdaBoost (i.e., 0.30, 0.13, 0.37, 0.97). This was due to the fact that XGBoost is based on more regularized models and better at avoid fitting data. In terms of the maximum error, AdaBoost slightly outperformed XGBoost (i.e., slightly less variance) but at the cost of significantly more computational time. Because the maximum Vicat prediction errors incurred by both XGBoost and AdaBoost were less than 2.9 (given by the team of expert chemists in the testing laboratory in the largest chemical factor in Thailand), chemical laboratories could depend on the predicted values instead of conducting time- and labor- intensive tests.

Nevertheless, for the initial and D2 datasets, GBM outperformed the others, meaning that it could tolerate inefficient feature selections more than the others did. In other words, XGBoost and AdaBoost are more sensitive to the number of features; they yield more errors in situations where the information is insufficient (i.e., inadequate relevant features) and where there exist too much noises (i.e., too many irrelevant features). This result has left us with opportunities to explore in the future research.

In terms of computational time, XGBoost was the fastest at 40.25 s, followed by GBM at 103.45 s, and AdaBoost was the slowest at 252.83 s. Overall, XGBoost consistently outshined GBM and AdaBoost in most of the metrics, while also being more computationally efficient.

Methods		GBM			XGBoost			AdaBoost		
Number of Features		41	61	71	41	61	71	41	61	71
Metrics	MAE	0.37	0.38	0.37	0.44	**0.26**	0.41	0.36	0.30	0.40
	MSE	0.32	0.29	0.31	0.45	**0.11**	0.46	0.35	0.13	0.41
	RMSE	0.57	0.54	0.55	0.67	**0.33**	0.68	0.59	0.37	0.64
	R^2	0.92	0.93	0.93	0.89	**0.98**	0.89	0.91	0.97	0.90
	Max Error	2.48	1.61	2.20	3.20	0.76	3.25	2.87	**0.73**	2.90
	% Max Error	2.65	1.72	2.35	3.41	0.81	3.47	3.06	**0.78**	3.09
	Processing time (s)	97.81	103.45	105.36	**37.78**	40.25	39.98	249.29	252.83	258.58

Fig. 4. Experimental results.

According to our best-performing model (XGBoost, with the following hyperparameter values: n_estimators = 50, max_depth = 7, min_child_weight = 1.1, and learning_rate = 0.51), we identified the top 10 features out of 61. Listed in descending order of importance scores, these features were Density, MFR, Gloss, Elongation, L-value, COF-Kinetic, Blown film, Haze, TC, and Tensile.

For future research, we plan to compare more diverse machine learning models to evaluate their efficacy of predicting different polymers and other crucial properties. Each model behavior would be studied in more detail to enhance understanding and increase trust particularly for industries.

5 Conclusion

In this study, we compared GBM, AdaBoost, and XGBoost algorithms, used to predict the Vicat softening point of LDPE, which is a widely used polymer in various industrial sectors. The research highlights the effectiveness of XGBoost in predicting LDPE parameters.

The methodology involved data preparation, model selection, and evaluation. The results from the experiments showed that XGBoost had superior predictive performance compared to AdaBoost and GBM because XGBoost is based on more regularized models and better at avoid fitting data. This was evident in the lower error metrics and higher R^2 values. Furthermore, the top 10 features influencing the prediction of LDPE's Vicat softening temperature were identified.

Overall, this study demonstrated the potential of machine learning algorithms like XGBoost in predicting important polymer properties, paving the way for more cost-efficient and sustainable material development.

Acknowledgments. We express our gratitude to the team of expert chemists from the largest plastic manufacturing facility for providing invaluable data on the maximum tolerable numerical errors in predicting Vicat softening temperature, as well as for their insightful review of the analysis of the experimental results (especially, the list of top 10 important features).

References

1. Polyethylene (Low Density) LDPE. https://www.bpf.co.uk/plastipedia/polymers/LDPE.aspx. Accessed 12 Nov 2023
2. Ledwani, D., Thakur, I., Bhatnagar, V.: Comparative analysis of prediction models for melt flow rate of C2 and C3 polymers synthesized using nanocatalysts. NanoWorld J. **8**(S1), S123–S127 (2022)
3. Li, H.B., Li, Q., Yan, M.L.: Influence of operation procedures on vicat softening temperature of thermoplastic materials. Adv. Mater. Res. **291–294**, 1820–1824 (2011)
4. Silfiani, M., Aprillia, H., Fitriani, Y.: Comparing various combined techniques at seasonal autoregressive integrated moving average (SARIMA) for electrical load forecasting. In: 2023 International Seminar on Intelligent Technology and Its Applications, Surabaya, Indonesia, pp. 376–381. IEEE (2023)
5. Shah, I., Akbar, S., Saba, T., Ali, S., Rehman, A.: Short-term forecasting for the electricity spot prices with extreme values treatment. IEEE Access **9**, 105451–105462 (2021)
6. Wu, P., Zhang, G., Li, Y., Chen, X.: Research on E-commerce inventory demand forecasting based on NAR neural network. Open Access Libr. J. **10**(5), 1–11 (2023)
7. Alam, M.S., Murshed, M., Manigandan, P., Pachiyappan, D., Abduvaxitovna, S.Z.: Forecasting oil, coal, and natural gas prices in the pre-and post-COVID scenarios: contextual evidence from India using time series forecasting tools. Resour. Policy **81**, 103342 (2023)
8. Li, S., Lu, X., Ouyang, J., Zhou, Y., Zhang, W.: K-means clustering algorithm and LSTM based short-term load forecasting for distribution transformer. In: 5th Asia Energy and Electrical Engineering Symposium (AEEES), Chengdu, China, pp. 1152–1156. IEEE (2023)
9. Shrivastava, V.K., Shrivastava, A., Sharma, N., Mohanty, S.N., Pattanaik, C.R.: Deep learning model for temperature prediction: a case study in New Delhi. J. Forecast. **42**(6), 1445–1460 (2023)
10. Xu, H., Wang, C.Y.: Demand prediction of chain supermarkets based on LSTM neural network. China Logist. Purchasing **3**, 42–43 (2021)
11. Vithitsoontorn, C., Chongstitvatana, P.: Demand forecasting in production planning for dairy products using machine learning and statistical method. In: 2022 International Electrical Engineering Congress (iEECON), Khon Kaen, Thailand, pp. 1–4. IEEE (2022)
12. Taufiqurrahman, A., Putrada, A.G., Dawani, F.: Decision tree regression with Adaboost ensemble learning for water temperature forecasting in aquaponic ecosystem. In: 6th International Conference on Interactive Digital Media (ICIDM), Bundung, Indonesia, pp. 1–5. IEEE (2020)
13. Han, J., et al.: Prediction of winter wheat yield based on multi-source data and machine learning in China. Remote Sens. **12**(2), 236 (2020)
14. Khoi, D.N., Quan, N.T., Linh, D.Q., Nhi, P.T.T., Thuy, N.T.D.: Using machine learning models for predicting the water quality index in the La Buong River Vietnam. Water **14**(10), 1552 (2022)
15. Baržić, M., Munitić, N.F., Bronić, F., Jelić, L., Lešić, V.: Forecasting sales in retail with XGBoost and iterated multi-step ahead method. In: 4th International Conference on Smart Systems and Technologies (SST), Osijek, Croatia, pp. 153–158. IEEE (2022)

16. Anggraeni, F., Adytia, D., Ramadhan, A.W.: Forecasting of wave height time series using AdaBoost and XGBoost, case study in Pangandaran, Indonesia. In: 4th International Conference on Data Science and Its Applications (ICoDSA), Bundung, Indonesia, pp. 97–101. IEEE (2021)
17. Udo, W., Muhammad, Y.: Data-driven predictive maintenance of wind turbine based on SCADA data. IEEE Access **9**, 162370–162388 (2021)
18. Gumus, M., Kiran, M.S.: Crude oil price forecasting using XGBoost. In: 2nd International Conference on Computer Science and Engineering (UBMK), Antalya, Turkey, pp. 1100–1103. IEEE (2017)
19. JayaBharath, M., Choudary, N.L., Pranay, C.S., Praveenya, M.D., Reddy, B.R.: An analysis of software maintainability prediction using ensemble learning algorithms. In: 3rd International Conference on Artificial Intelligence and Signal Processing (AISP), Vijayawada, India, pp. 1–5. IEEE (2023)

SDN Enabled L2 Switch Implementation and Its Performance Evaluation Through P4 Programming

Sakar Pudasaini[1], Babu R. Dawadi[1(✉)], Roshani Ghimire[2], Giovanni Pau[3], and Binod Sapkota[1]

[1] Institute of Engineering, Pulchowk Campus, Tribuvan University, Kathmandu, Nepal
078msice015.sakar@pcampus.edu.np, baburd@ioe.edu.np, replybinod@gmail.com
[2] Advanced College of Engineering and Management, Tribhuvan University, Kathmandu, Nepal
roshanigd@gmail.com
[3] Faculty of Engineering and Architecture, Kore University of Enna, Enna, Italy
giovanni.pau@unikore.it

Abstract. The concept of SDN (Software-Defined Networking) originally came with the purpose of disaggregating control plane and data plane. However, current scope of SDN has surpassed the original goal. It sets out to introduce programmability in the control plane for better visibility and manageability of networking devices. The idea has been recently extended to program the data plane. In the present scenario P4 (Programming Protocol-Independent Packet Processors) stands as the de-facto language to program the data plane. In this study, we implemented an L2 (Layer 2) switch and evaluated it performance using P4 programming language. The performance of the switch was evaluated in terms of throughput (Mbps) and latency (ms). The throughput was found to be nominal as the switch was implemented in software and the latency for first packet was also high since the switch's table was empty. However, once the table was filled, subsequent packets suffered very low latency.

Keywords: SDN · Programmable Data Plane · P4 · Data Plane · Control Plane · Network Programmability

1 Introduction

The original intent of SDN was to disaggregate control plane and data plane from individual networking devices and instead have a logically centralized control plane which would then control data plane of multiple devices. This idea proved to be a revolutionary concept as it unlocked a range of possibility. With fixed-function ASICs (Application-Specific Integrated Circuit) switches, network

Funded by UGC Nepal under Collaborative Research Grants ID: CRG-078/79-Engg-01.

P. Meesad et al. (Eds.): IC2IT 2024, LNNS 973, pp. 33–42, 2024.
https://doi.org/10.1007/978-3-031-58561-6_4

programmability was limited to the control plane only. Northbound APIs were used by applications for network telemetry, dictating network behaviour etc. The idea of network programmability penetrated deeper with the introduction of programmable switches where even the data plane could be programmed which lets the operator implement any protocol rapidly.

P4 is a programming language to program the data plane. It is a domain-specific language with a number of constructs optimized for network data forwarding. The goal of P4 is Target independence, Protocol independence and Reconfigurability. Currently, P4 is the de facto language to program the data plane. P4 can be used to program the programmable switches with existing protocols such as Ethernet, IP, TCP etc. or it can be used to program a custom protocol as per user's requirement.

Deployment of new protocols in the data plane has always been a challenge due to fixed-function ASIC switches. The current standard way of implementing network infrastructure involves a bottom-up design where the hardware defines the network functions, thus network functionality is limited and depended on the underlying ASIC. The only possible way to introduce new protocol would be to implement it in software at the expense of increased CPU load, latency and reduced throughput. With programmable data plane and P4, however, it is possible to use a more beneficial approach of top-down design where the network is instead designed by what is wanted or needed by the developers rather than being limited by the hardware.

A fully functioning layer 2 SDN switch was implemented and evaluated using P4 in this research. A layer 2 switch is a networking device which forwards a frame from one interface to another on the basis of layer 2 headers such as Ethernet. An ethernet header consists of three main fields namely, Destination MAC, Source MAC, Ethertype. A layer 2 switch makes its forwarding decision based on the destination MAC on the header and its MAC address table. Whenever a frame is received, the switch inspects the destination MAC on the frame, performs a lookup on its MAC address table to find which egress interface should be selected for forwarding the frame and finally forwards the frame through the egress interface. The job of populating MAC address table is carried out by the control plane. A layer 2 SDN switch was implemented to demonstrate the capability of programmable data plane using P4 to describe the elements of ethernet header as well as the control plane logic to forward the frame from ingress interface to egress interface. Finally, an evaluation of the implemented switch was carried out to measure the performance of the implemented switch. P4 programming can be implemented in hardware or in software. Due to high cost associated with hardware, this study was limited to emulation using Mininet.

This study demonstrated that Programmable Data Plane along with P4 can be utilized to implement user defined data plane protocol as well as control plane logic in a relatively rapid and target agnostic manner enabling future development in the Programmable Data Plane domain. This paper is divided into 5 sections. Section 1 presents the introduction to SDN and data plane programmability, Sect. 2 discusses about the necessary background and works that has been carried out in data plane programmability, Sect. 3 presents the methodology to carry out the research work, Sect. 4 presents the results obtained from the work and finally Sect. 5 briefly discusses about the conclusion.

2 Background and Related Work

In SDN, the control plane is physically separated from the forwarding plane, and one control plane controls multiple forwarding devices. OpenFlow is commonly used to control different data plane devices in a vendor agnostic way [1]. OpenFlow matches packets on predefined header fields and has scaling issues. There are three main goals of P4 are Reconfigurability (change data plane logic anytime), Protocol Independence (any arbitary protocol can be implemented) and Target Independence (can be compiled against any hardware).

Software-Defined Networking: SDN is an approach to implement networks. Instead of having distributed control plane, SDN utilizes logically centralized control plane with distributed data plane. Control Plane refers to any action that controls the data plane such as creating tables used by data plane e.g. routing table, MAC table etc. Data Plane on the other hand is responsible for implementing the behavior dictated by control plane on individual packets and actually forwarding the data packets [2].

P4 Switch Workflow: $P4_{16}$ is the latest specification of the language and supports multiple different targets and pipeline architectures [3]. The compiler compiles the user created target-independent P4 source code to target specific binary. It generates two important files after processing the source code. First, it generates a target specific data plane configuration. This configures the parser, match-action stages and deparser of the target. Second, it generates a runtime API that is used by control plane to interact with the data plane. It contains the information needed to manipulate tables in the data plane [4].

Data Plane to Control Plane Communication: The data plane could notify the control plane in two ways. It could either do packet-in where the switch sends a copy of the original packet along with the metadata to the controller via a specially reserved port as shown in Fig. 1 or it could only send a packet digest [5]. Packet digests are much smaller than the original packet and only contain subset of the headers and/or P4 metadata associated with the packet [6] as shown in Fig. 2. The controller computes the digests and communicates with the data plane using runtime APIs [4]. The P4Runtime is an API for Control Plane to interact with Data Plane. From an SDN architecture perspective, it can be thought of a Southbound API. It utilizes gRPC for communication and protobuf for message serialization [6]. Similar to OpenFlow, P4Runtime provides a simple, high-level API for managing the forwarding behavior of a switch like adding or removing entries to a table.

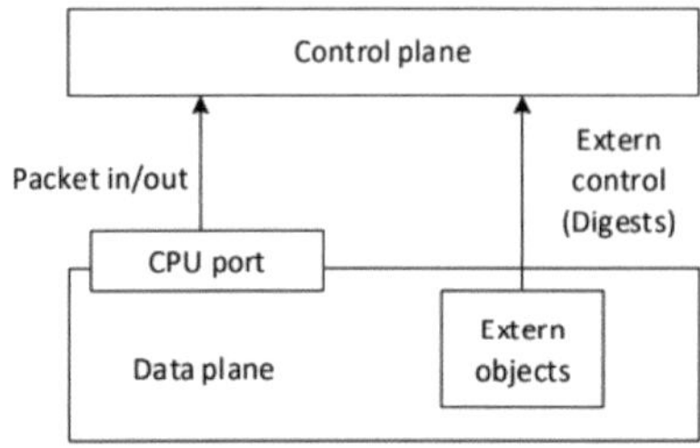

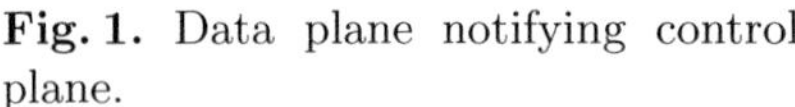

Fig. 1. Data plane notifying control plane.

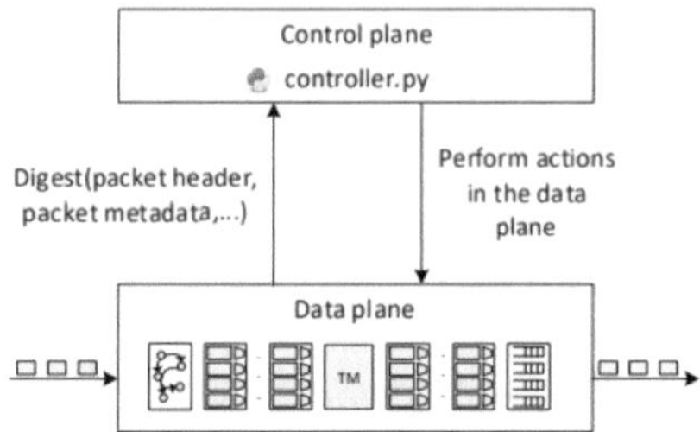

Fig. 2. Digest based learning.

V1Model and BMv2: The V1model architecture is shown in Fig. 3. It has a parser, match-action pipeline and a deparser as Programmable Component and Traffic Manager as Configurable (Fixed-Function) Component [7].

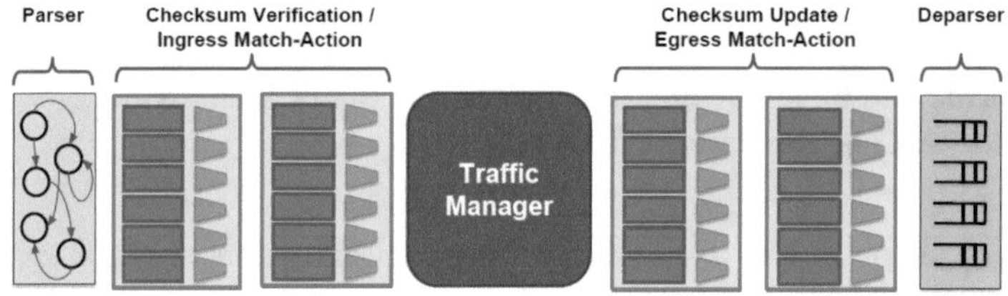

Fig. 3. V1Model architecture.

The Behavioral Model version 2 (BMv2) is a reference P4 software switch implementation. It takes a JSON file generated from P4 Compiler as an input to configure its forwarding plane logic [8]. It is an abstract switch model and can utilize many architecture such as v1model. A BMv2 target utilizing v1model is referred as simple_switch [9].

P4 and Programmable Data Plane has been used for stateful Layer 4 Load Balancing, where hundreds of software load balancer can be replaced with a single modern switching ASIC [10]. P4 is also studied to implement network security within data plane to mitigate large-scale attacks with low overhead [3]. Another notable and high value work is In-band Network Telemetry which gives more fine-grained telemetry information directly from the data plane without the involvement of control plane [11]. Apart from these work, P4 has been studied to implement custom protocols to gain better performance in terms of bandwidth, latency and jitter [12]. Detection of DDoS attack in P4-SDN is another domain of active research [13]. Access Control and Authentication with P4 is another topic of interest in the domain of Network Security which aims to mitigate MITM attacks [14]. The approach of implementing SDN switch presented in this study provides an advantage compared to OpenFlow approach in that the behaviour can be modified anytime and has less protocol overhead.

3 Methodology

Mininet VM was installed on top of VMware Workstation Pro hypervisor due to its better performance [15]. P4 Compiler and BMv2 Switch for P4 programming was setup inside this VM. An L2 switch was programmed using P4 programming. The compiled program was successfully emulated with BMv2 switch in mininet. Figure 4 shows the emulated network topology. It consists of a single controller, single switch and four hosts. The number of hosts can be adjusted as per requirements. However, due to the rudimentary implementation of the controller, only single switch is supported. To support multiple switches, the controller would be more complex and would also require to detect and block bridging loops.

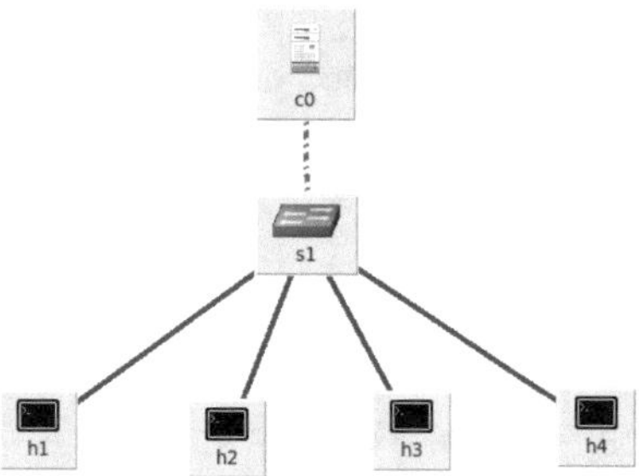

Fig. 4. Emulated network topology.

The experiment was conducted inside a VM in a host PC with CPU: Intel(R) Core(TM) i5-6300U CPU @ 2.40 GHz, RAM: 8 GB DDR4, and 256 GB NVMe SSD storage. To comply the implemented layer 2 switch with the definition of SDN, it was implemented as two separate components: Data Plane and Control Plane. Finally, similar topology was launched within Mininet using OpenFlow switch. The Southbound API was captured in both cases and their overhead was studied.

Data Plane: The Data Plane of the L2 switch was implemented using P4 programming language. The implemented data plane logic is presented in Fig. 5. Three tables were created in Ingress pipeline to perform MAC learning, frame forwarding and broadcasting. The packet first passes through the Source MAC table in the ingress pipeline, here the key to match on is source MAC address, if there is a match no action is taken and the packet is moved forward in the pipeline but if there is not a match, it would indicate that this is a new MAC address and needs to be learned. Thus, a digest message is sent to the controller using P4Runtime API to inform about this new MAC address. Next in the pipeline comes Destination MAC table, here the key to match on is destination MAC address to find out which egress port should be selected. If there is a match here an egress port is set and the frame is forwarded to the Traffic Manager but if there is a miss in this table, the packet is sent further in the forwarding pipeline to the broadcast table. The Broadcast table simply matches on the ingress port and sets mcast_grp standard metadata so, that the frame is broadcasted to every other port except the ingress port. The actual process of broadcasting is done by the Packet Replication Engine which looks at the mcast_grp and make appropriate number of copies and send to appropriate egress ports. The information about egress ports for each ingress port is provided from the control plane.

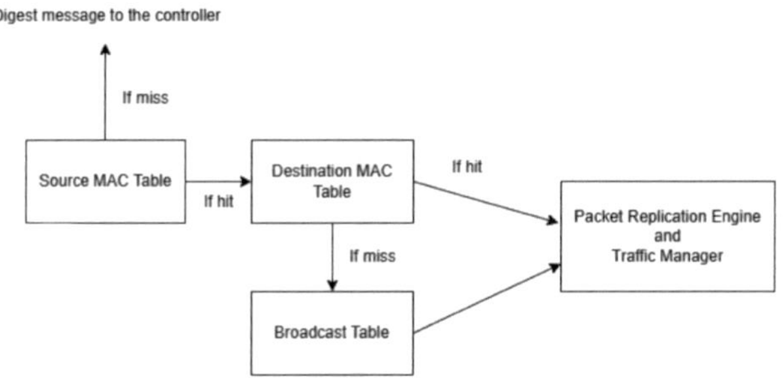

Fig. 5. Implemented layer 2 frame processing logic.

Control Plane: The Control Plane for the Switch was implemented in Python utilizing P4Runtime API for data plane - control plane communication. The control plane logic is presented in Fig. 6. The Controller acts as a gRPC client and connects to the data plane switch at TCP port 50051 which acts as a gRPC server. First, the controller sends Master Arbitration Update to establish itself as the controller with write permission next it configures the switch with Set-ForwardingPipelineConfig RPC call. The controller is also responsible for configuring the Packet Replication Engine of the V1Model for multicast support. The controller has been programmed to create separate multicast group for each ingress port, so that the BUM frames are replicated and broadcasted out to every other ports except the ingress port itself. Finally the controller constantly listens for the digest messages from the data plane and if the received MAC address is unseen before it Writes that information back to the table of the data plane switch.

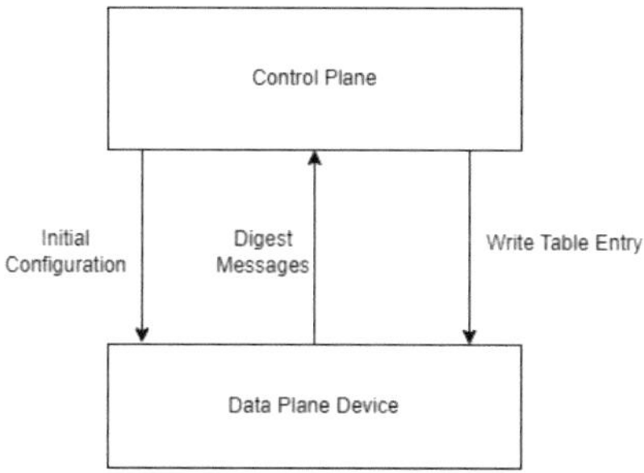

Fig. 6. Control plane functionality.

4 Results and Analysis

The L2 switch implemented with P4 was able to perform L2 forwarding. The hosts h1, h2, h3, and h4 were set on same L2 domain with IP address from the same subnet and were able to ping each other as presented in Fig. 7.

A fully Software Defined Layer 2 switch was implemented whose data plane and control plane behaviour could be dictated by software entirely. The implemented switch and controller was capable of MAC address learning, frame forwarding as well as frame broadcasting.

```
mininet> h1 ping h2
PING 10.0.1.2 (10.0.1.2) 56(84) bytes of data.
64 bytes from 10.0.1.2: icmp_seq=1 ttl=64 time=9.85 ms
64 bytes from 10.0.1.2: icmp_seq=2 ttl=64 time=0.750 ms
64 bytes from 10.0.1.2: icmp_seq=3 ttl=64 time=0.737 ms
64 bytes from 10.0.1.2: icmp_seq=4 ttl=64 time=0.712 ms
64 bytes from 10.0.1.2: icmp_seq=5 ttl=64 time=1.67 ms
64 bytes from 10.0.1.2: icmp_seq=6 ttl=64 time=0.724 ms
64 bytes from 10.0.1.2: icmp_seq=7 ttl=64 time=1.56 ms
64 bytes from 10.0.1.2: icmp_seq=8 ttl=64 time=0.755 ms
64 bytes from 10.0.1.2: icmp_seq=9 ttl=64 time=1.78 ms
^C
--- 10.0.1.2 ping statistics ---
9 packets transmitted, 9 received, 0% packet loss, time 8115ms
rtt min/avg/max/mdev = 0.712/2.060/9.854/2.788 ms
```

Fig. 7. Hosts H1 and H2 connectivity test.

```
> Frame 16: 14033 bytes on wire (112264 bits), 14033 bytes captured (112264 bits) on interface lo, i
> Ethernet II, Src: 00:00:00_00:00:00 (00:00:00:00:00:00), Dst: 00:00:00_00:00:00 (00:00:00:00:00:0
> Internet Protocol Version 4, Src: 127.0.0.1, Dst: 127.0.0.1
> Transmission Control Protocol, Src Port: 49332, Dst Port: 50051, Seq: 446, Ack: 266, Len: 13967
> HyperText Transfer Protocol 2
> HyperText Transfer Protocol 2
> HyperText Transfer Protocol 2
  GRPC Message: /p4.v1.P4Runtime/SetForwardingPipelineConfig, Request
v Protocol Buffers: /p4.v1.P4Runtime/SetForwardingPipelineConfig,request
  v Message: <UNKNOWN> Message Type
    > Field(3):
    > Field(4): 3 (uint32)
    > Field(5):
> HyperText Transfer Protocol 2
```

Fig. 8. Packet capture of P4Runtime.

Figure 8 shows a capture of P4Runtime API. Specifically SetForwarding-PipelineConfig API can be observed, this API is made from the controller to the switch to load the data plane configuration. Figure 9 shows an API call from the client (controller) to the server (switch) to write an table entry to the data plane. It can be seen that the message is compressed with binary encoding and cannot be interpreted directly, this is due to the use of protocol buffer in gRPC. The message only contained the ingress port number and the source MAC address.

Figure 10 shows OpenFlow capture for the same scenario. Here, we can observe that the OpenFlow PACKET_IN message contains the whole of the ethernet header as well as the ARP header. Thus, the implemented P4Runtime

```
> GRPC Message: /p4.v1.P4Runtime/Write, Request
v Protocol Buffers: /p4.v1.P4Runtime/Write,request
  v Message: <UNKNOWN> Message Type
    v Field(3):
        [Field Name: <UNKNOWN>]
        .001 1... = Field Number: 3
        .... .010 = Wire Type: Length-delimited (2)
        Value Length: 2
        Value: 1001
    v Field(4):
        [Field Name: <UNKNOWN>]
        .010 0... = Field Number: 4
        .... .010 = Wire Type: Length-delimited (2)
        Value Length: 34
        Value: 0801121e121c08fd8bd011120c080112080a060800000000001111a070a0508b7b6910a
> HyperText Transfer Protocol 2
```

Fig. 9. P4Runtime write API.

```
> Frame 24: 126 bytes on wire (1008 bits), 126 bytes captured (1008 bits) on interface lo, id 0
> Ethernet II, Src: 00:00:00_00:00:00 (00:00:00:00:00:00), Dst: 00:00:00_00:00:00 (00:00:00:00:00:00
> Internet Protocol Version 4, Src: 127.0.0.1, Dst: 127.0.0.1
> Transmission Control Protocol, Src Port: 58302, Dst Port: 6653, Seq: 201, Ack: 45, Len: 60
v OpenFlow 1.0
    .000 0001 = Version: 1.0 (0x01)
    Type: OFPT_PACKET_IN (10)
    Length: 60
    Transaction ID: 0
    Buffer Id: 0xffffffff
    Total length: 42
    In port: 1
    Reason: No matching flow (table-miss flow entry) (0)
    Pad: 00
  > Ethernet II, Src: de:77:e3:62:62:ee (de:77:e3:62:62:ee), Dst: Broadcast (ff:ff:ff:ff:ff:ff)
  > Address Resolution Protocol (request)
```

Fig. 10. OpenFlow capture.

MAC learning procedure is more efficient than MAC learning in OpenFlow switches as only the required part of the header and metadata were sent to the controller for learning purpose, whereas in OpenFlow complete header is sent to controller as Packet In message.

For the performance analysis ICMP (Internet Control Message Protocol) packets were sent to measure the average latency. It was observed that the first ping from a host to another took longer than the subsequent pings since, it has to go through the ARP resolution process and the table entries would also be unavailable at that moment, forcing the switch to replicate the packets and broadcast. Throughput was another parameter that was measured. iperf was used to measure the throughput between the hosts. For each parameter 30 measurements were made.

The average RTT (Round Trip Time) for first packet was found to be around 15.5 ms, while the latency for subsequent packets were found to be at around 0.876 ms. Similarly, the bandwidth measured between the hosts using iperf averaged at around 26.4 Mbps.

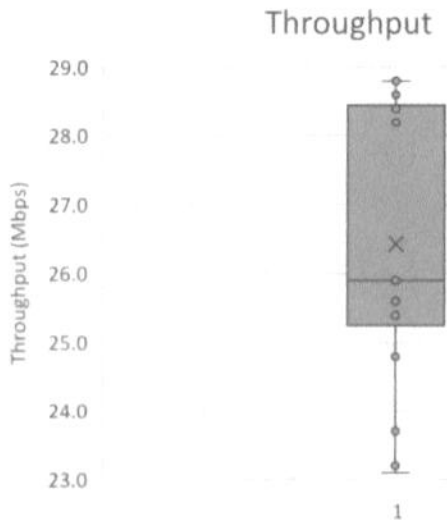

Fig. 11. Measured throughput.

Figure 11 shows the box plot for the measured throughput between two hosts. The achieved throughput varied from 23.1 to 28.8 Mbps with a median of 25.9 Mbps for our system setup. Similarly, Fig. 12a and 12b shows the latency measurements between two hosts when the destination MAC is unpopulated in the table entries and when the destination MAC is already populated in the table entries respectively. It can be observed that the latency in the case of unpopulated destination MAC address is significantly higher than in the case when destination MAC address is already populated. In the case of unpopulated destination MAC the latency varied from 12.7 to 18 ms whereas in the case of populated destination MAC the latency varied from 0.616 to 1.6 ms. One of the reason for such low value of latency in this case was due to hosts being emulated within the same system.

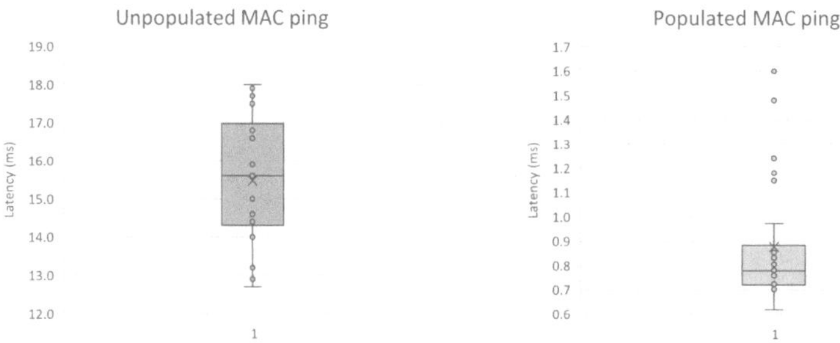

(a) Measured Latency when destination MAC is unpopulated.

(b) Measured Latency when destination MAC is populated.

Fig. 12. Comparison of latency.

5 Conclusion and Future Works

The implementation of an SDN enabled Layer 2 switch in P4 and a corresponding controller in Python using the P4Runtime API has been successfully achieved. The study involved designing the P4 code for the Layer 2 switch and the Python code for the controller, which communicates with the switch using the P4Runtime API. The Layer 2 switch implemented essential features such as MAC learning, forwarding and broadcasting, while the controller provided centralized control point for the switch. Throughout the study, several challenges were encountered, including the need to understand the complex P4 language and its interaction with the P4Runtime API. However, with proper research and testing, these challenges were successfully overcome. This type of study can be applied in developing more complex SDN applications in the future.

The implemented Layer 2 switch has a baseline feature of learning, forwarding and broadcasting and the controller implemented has basic capability of managing only a single switch. In the future, more features can be added within the data plane such as MAC address filtering and VLAN separation. Similarly, the controller can be made more robust by adding the capability to handle multiple switches and make topology-aware to avoid bridging loops.

Acknowledgements. This research is supported by University Grants Commission (UGC) Nepal Collaborative Research Grants (ID: CRG-078/79-Engg-01).

References

1. Bosshart, P., et al.: P4: programming protocol independent packet processors. ACM SIGCOMM Comput. Commun. Rev. **44**(3), 87–95 (2014)
2. Edgeworth, B., Rios, R.G., Hucaby, D., Gooley, J.: CCNP and CCIE Enterprise Core ENCOR 350-401 Official Cert Guide. Cisco Press (2019)
3. Gao, Y., Wang, Z.: A review of P4 programmable data planes for network security. Mob. Inf. Syst. (2021)
4. University of South Carolina Cyberinfrastructure (SC CI): Introduction to P4 Programmable Data Planes. http://ce.sc.edu/cyberinfra/workshops/Material/P4-BMv2.pdf. Accessed 08 Feb 2023
5. NSG Group, ETH Zurich: BMv2 Simple Switch. https://github.com/nsg-ethz/p4-learning/wiki/BMv2-Simple-Switch. Accessed 08 Feb 2023
6. P4 API Working Group: P4runtime specification. https://p4.org/p4-spec/p4runtime/main/P4Runtime-Spec.html. Accessed 08 Feb 2023
7. Gurevich, V., Fingerhut, A.: P416 programming for intel Tofino using intel P4 studio. https://opennetworking.org/wp-content/uploads/2021/05/2021-P4-WS-Vladimir-Gurevich-Slides.pdf. Accessed 08 Feb 2023
8. P4 Language Consortium: P4 tutorial. https://github.com/p4lang/tutorials/blob/master/P4_tutorial.pdf. Accessed 08 Feb 2023
9. P4 Language Consortium: Behavioral model. https://github.com/p4lang/behavioral-model/blob/main/README.md. Accessed 08 Feb 2023
10. Miao, R., Zeng, H., Kim, C., Lee, J., Yu, M.: SilkRoad: making stateful layer-4 load balancing fast and cheap using switching ASICs. In: Proceedings of the Conference of the ACM Special Interest Group on Data Communication, pp. 15–28 (2017)
11. Joshi, M.: Implementation and evaluation of in-band network telemetry in P4. Master's thesis, KTH Royal Institute of Technology (2021)
12. Claeson, O., Kruse, W.: High-speed, low-latency, and secure networking with P4. Master's thesis, Chalmers University of Technology (2021)
13. Lin, T., Wang, C., Tuan, Y., Tsai, M., Chen, Y.: A study on traffic asymmetry for detecting DDoS attack in P4-based SDN. J. Inf. Sci. Eng. **38**(6), 1265–1283 (2022). https://doi.org/10.6688/JISE.202211_38(6).0009
14. Bhattacharya, A., Rana, R., Datta, S., Venkanna, U.: P4-sKnock: a two level host authentication and access control mechanism in P4 based SDN. In: 2022 27th Asia Pacific Conference on Communications (APCC), pp. 278–283 (2022). https://doi.org/10.1109/APCC55198.2022.9943765
15. Mininet Project: Mininet VM setup notes. http://mininet.org/vm-setup-notes/. Accessed 08 Feb 2023

Formal Verification of Sequence Diagram with State Invariants Using Timed Automata

Supapitch S. Thitareedechakul(✉) and Wiwat Vatanawood

Department of Computer Engineering, Chulalongkorn University, Bangkok, Thailand
6470288521@student.chula.ac.th, wiwat@chula.ac.th

Abstract. In a design model, the UML sequence diagrams typically play a crucial role in explaining the behaviors of the software system by showing the interactions between their objects. Especially, a sequence diagram with state invariants intentionally shows the appropriate runtime states of the objects at the particular points of time in their timelines. In practice, state-dependent behavior for an object is provided. An object responds differently to the same event depending on what state it is in. Moreover, due to the possibility of the hybrid uses of the synchronous and asynchronous interactions in this diagram, the nondeterministic situation may be unfortunately introduced, causing more complex and difficult to verify the final design. In this paper, we propose a formal verification method to ensure the basic properties of the design model drawn in these sequence diagrams with state invariants. A set of transformation rules are proposed to convert the original sequence diagram into its corresponding timed automata. The behaviors of the resulting timed automata would be simulated and verified using UPPAAL tool. Moreover, the behavioral properties, including deadlock, liveness, and safeness, of the design model are also formally verified using timed computation tree logic.

Keywords: Sequence Diagram with State Invariants · Timed Automata · Formal Verification

1 Introduction

In the typical software development life cycle [1], the software design phase is responsible for translating the requirements specification into design diagrams. The sequence diagram is one of the essential diagrams that plays a crucial role in explaining the behavior of the object interactions within the software system. However, in large and complex software designs, it can be difficult to ensure the correctness, safety, liveness, and completeness of the design diagrams manually. Therefore, the formal verification approach can be exploited to effectively cope with both synchronous and asynchronous messaging, as well as the parallel behavior processes shown in the sequence diagrams. In UML 2.5, the sequence diagram with state invariants (SDSI) is introduced [2–4] to provide the state invariants at runtime mode. This enables the software designer to visualize the state-dependent behavior of the software system, especially for the event-driven architecture. Several researchers proposed a formal verification approach to ensure the basic

P. Meesad et al. (Eds.): IC2IT 2024, LNNS 973, pp. 43–54, 2024.
https://doi.org/10.1007/978-3-031-58561-6_5

properties of the sequence diagram. For instance, [5] generally proposed the formal model of the behaviors, but neither the synchronous and asynchronous messages, nor parallel processes were discussed.

In this paper, we propose an alternative approach to do formal verification of the basic properties of SDSI using model checking. We propose a set of transformation rules to map the original SDSI into timed automata [6]. The UPPAAL simulation tool [7] is exploited to perform the model checking of the resulting timed automata using timed computation tree logic (TCTL). The remainder of this paper is organized as follows: Sect. 2 describes the background of SDSI, timed automata, and the UPPAAL tool. Section 3 describes our transformation approach to map SDSI into the corresponding timed automata. Section 4 presents our case study, and Sect. 5 concludes this paper.

2 Background

2.1 Sequence Diagram with State Invariants

Sequence Diagram with State Invariants (SDSI) is a diagram that illustrates the number of participants, their interactions, and their behaviors. Within this diagram, participants will have state invariants for a specific period of time in their lifecycle.

In general, the Sequence diagram was modeled in UML 2.0 [2]. In the diagram, which consists of participant symbols and their names, participants are placed in a vertical alignment, and a dashed line is drawn to indicate their lifeline and sequences of processes in the system.

When participants start the process, activation bars are drawn, and arrows and messages are placed between the requester and the provider. The requester will send a message to the provider, and when the message is received from the requester, the provider will use the message to start processes in their systems. In general, arrow messages may include both synchronous and asynchronous messages.

The Sequence diagram includes sequence fragments to reduce complexity when designing it. These fragments are designed for various behaviors such as "alt," "par," "loop," "break," "region," and more.

In UML 2.5 [4], the Sequence diagram defines state invariants using symbols that represent interaction fragments. These symbols are typically represented in rounded boxes and placed on participants lifelines. They can also include assigned messages within the rounded box, such as "start". State invariants symbols indicate a specific period of time during which participants will have a certain state. an example of sequence diagram symbol is shown in Fig. 1.

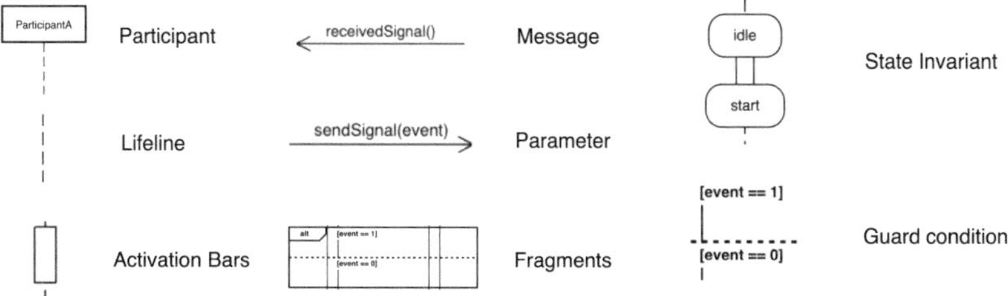

Fig. 1. Sequence diagram symbols [2, 4].

2.2 Timed Automata

From the theory of timed automata, a timed automaton is a finite automaton [3] consisting of a finite set attached to a clock. It is used to simulate processes within systems and observe state transitions. In general, timed automata will have locations representing states in the system, starting from l_0, which is the initial location in the system. When the system undergoes a transition within a specific period of time, there must be constraints, acting as guards on a transition line, to indicate that the system will change from its current state to the next state in accordance with the constraints on those transitions. Finally, the system will transition to a finite state.

Timed automata can model the behavior of systems and are used to verify software requirements specifications that are hard and complex to verify manually. They can also automatically verify systems with nondeterministic behavior. An example of a timed automata is shown in Fig. 2.

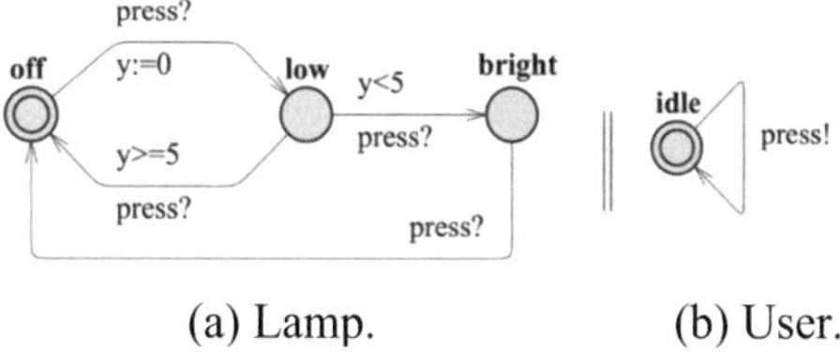

(a) Lamp. (b) User.

Fig. 2. Timed automata [6].

Definition 1: Timed automata is defined as a 6-tuple [6]

$$TA = (L, l_0, C, A, E, I)$$

- L is a set of locations
- $l_0 \in L$ is an initial location
- C is a set of clocks
- A is a set of actions
- $E \subseteq L \times A \times B(C) \times 2^c \times L$ is a set of edges or transitions between locations with an action, a guard, and a set of clocks to be reset or changed.
- $I : L \to B(C)$ is assigned invariants to locations

2.3 UPPAAL Model Checking

UPPAAL [8] is a model checking tool designed to support real-time system simulation and verification. It was founded and developed by Uppsala University and Aalborg University, with the first version released in 1995 [7].

The UPPAAL tool operates based on the theory of timed automata [6]. In this framework, locations represent states within specific time periods in the system, edges represent transitions between these locations, and clock variables are used to define constraints as guards on these edges. UPPAAL can be used to model the processes within a system and verify that these processes can reach every location or state.

3 Our Transformation Approach

In our transformation approach, our SDSI is defined and the transformation rules are defined to guide the mapping our SDSI into timed automata.

3.1 Define Sequence Diagram with State Invariants

Definition 2: SDSI is defined as a 7-tuple

$$SDSI = (P, SI, LL, M, F, GC, PR)$$

- P is a set of participants
- SI is a set of state invariants
- LL is a set of lifelines
- M is a set of messages
- F is a set of sequence fragments
- GC is a set of guard conditions
- PR is a set of parameters

When comparing the SDSI shown in Fig. 3(a) to Definition 2, we can explain that Participant1 is represented as $p_1 \in P$, and Participant2 as $p_2 \in P$, both of which are subsets of participants. State invariants, such as "idle" represented by $si_1 \in SI$, "sent" by $si_2 \in SI$, and "start" by $si_3 \in SI$, are also subsets of state invariants. The top box and dashed line associated with p_1 are denoted as $ll_1 \in LL$, while those associated with p_2 are denoted as $ll_2 \in LL$, , both of which are subsets of lifelines. The arrow and messages, such as "sendSignal(event)", placed between p_1 and p_2 are collectively referred to as $m_1 \in M$, a subset of messages. Similarly, the sequence fragment "alt" placed between p_1 and p_2 is represented as $f_1 \in F$, which is a subset of sequence fragments. The guard conditions under the sequence fragment, such as "event == 1", are identified as $gc_1 \in GC$, and are subsets of guard conditions. Lastly, a parameter assigned to the message "sendSignal(event)" is denoted as $pr_1 \in PR$, and is a subset of parameters.

3.2 Compare and Analyze the Structure of SDSI and Timed Automata

We model the SDSI as shown in Fig. 3(a), to simulate the behavior of the system. In the system, there are two participants, ParticipantA and ParticipantB. The timeline of events is as follows:

In step 1, ParticipantA is in the state "idle," and ParticipantB is in the state "off."
In step 2, ParticipantA sends a message "sendSignal(event)" to ParticipantB. ParticipantA transitions to the "send" state, while ParticipantB transitions to the "on" state.
In step 3, after ParticipantB receives a message from ParticipantA, two possible scenarios exist:

> If the parameter "event" is equal to 1, ParticipantB sends a message "receivedSignal()" back to ParticipantA. ParticipantB transitions to the "successful" state, while ParticipantA transitions to the "start" state.
> If the parameter is equal to 0, ParticipantB sends a message "closedSignal()" back to ParticipantA. ParticipantB transitions to the "unsuccessful" state, while ParticipantA transitions to the "stop" state.

We analyzed the structure of the SDSI and compared it with timed automata, modeling it in the UPPAAL tool. The results of the timed automata modeling are shown in Fig. 3(b).

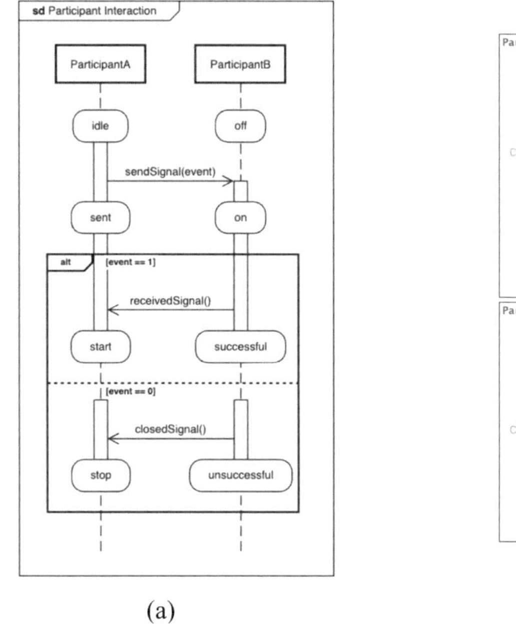

(a)

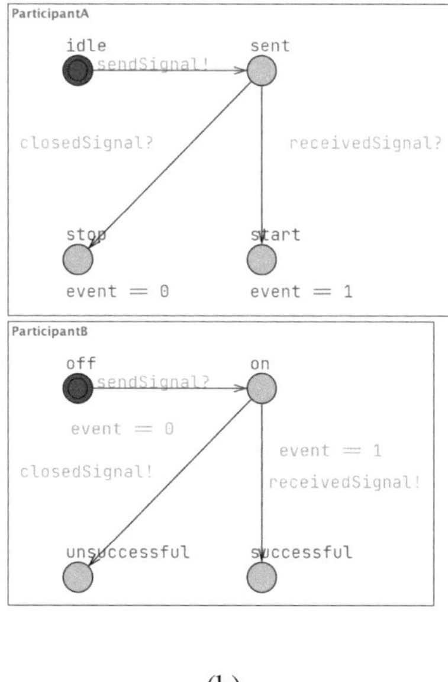

(b)

Fig. 3. An example of SDSI (a); The results of the timed automata (b).

3.3 Define Transformation Rules for Mapping SDSI into Timed Automata

After comparing and analyzing the structure of SDSI and timed automata, we can add the "Variables" element to Definition 1, which is

Definition 3: TAV is defined as a 7-tuple

$$TAV = (L, l_0, C, A, E, I, V)$$

- L is a set of locations
- $l_0 \in L$ is an initial location
- C is a set of clocks
- A is a set of actions
- $E \subseteq L \times A \times B(C) \times 2^c \times L$ is a set of edges or transitions between locations with an action, a guard, and a set of clocks to be reset or changed.
- $I : L \rightarrow B(C)$ is assigned invariants to locations
- V is a set of variables

We can define transformation rules for mapping SDSI into timed automata.

Transformation Rule 1: Participant
Participants p_1 and p_2, where $p_1, p_2 \in P$, in SDSI can be modeled as ta_1 and ta_2, where $ta_1, ta_2 \in TAV$, in timed automata. Each participant will have a state invariant si_0 and transition to si_1, where $si_0, si_1 \in SI$, which can be compared to timed automata that have locations l_0 transitioning to l_1, where $l_0, l_1 \in L$. For the example shown in Fig. 4, p_1 and ta_1 are labeled as ParticipantA, p_2 and ta_2 are labeled as ParticipantB.

Transformation Rule 2: State Invariants
The state invariants in participant $p_1 \in P$, denoted as si_0 and si_1, where $si_0, si_1 \in SI$, in SDSI, can be modeled as locations l_0 and l_1, where $l_0, l_1 \in L$, in timed automata. This comparison can be extended to include state invariants in participants and locations in timed automata. For the example shown in Fig. 4, si_0 and l_0 are labeled as "idle", si_1 and l_1 are labeled as "start".

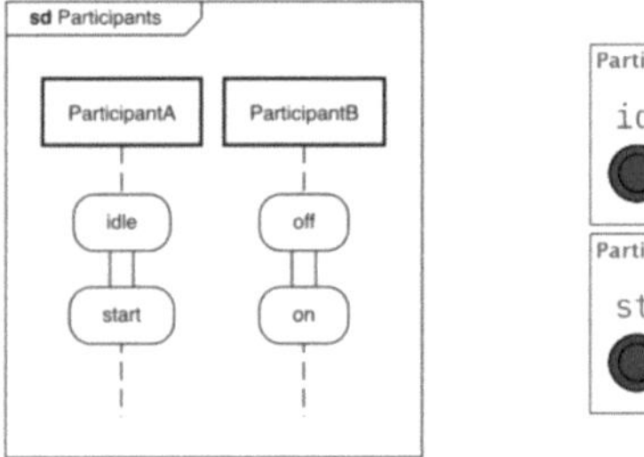

Fig. 4. Mapping the participants, the state invariants, and lifeline in SDSI into timed automata.

Transformation Rule 3: Lifeline
The lifeline of participant $p_1 \in P$, denoted as $ll_1 \in LL$, with state invariants $si_0 \rightarrow si_1$, where $si_0, si_1 \in SI$, in SDSI, can be modeled as a transition between locations l_0 and l_1, where $l_0, l_1 \in L$, in timed automata $ta_1 \in TAV$. This comparison can also be extended to include transitions of participants in SDSI and timed automata, as shown in Fig. 4.

Transformation Rule 4: Message
A message, denoted as $m_1 \in M$, sent from participant $p_1 \rightarrow p_2$, where $p_1, p_2 \in P$, in SDSI, can be modeled as channel synchronization at $a_1 \in A$ in both sender and receiver modes, synchronizing from $ta_1 \rightarrow ta_2$, where $ta_1, ta_2 \in TAV$. For the example shown in Fig. 5, m_1 is labeled as "sendSignal()" and a_1 in the sender mode is labeled as "sendSignal!", while in the receiver mode it is labeled as "sendSignal?".

Transformation Rule 5: Parameter
The parameter, denoted as $pr_1 \in PR$, is assigned in a message at $m_1 \in M$, sent from participant $p_1 \rightarrow p_2$, where $p_1, p_2 \in P$, in SDSI, can be modeled as a variable declaration at $v_1 \in V$, from $ta_1 \rightarrow ta_2$, where $ta_1, ta_2 \in TAV$. For the example shown in Fig. 5, pr_1 and v_1 are labeled as "event".

Transformation Rule 6: Guard Condition
A guard condition, denoted as $gc_1 \in GC$, is assigned within a fragment of participant $p_1 \in P$, in SDSI. If ta_1 is the sender, it can be modeled as a guard on the transition line at $e_1 \in E$, which represents a condition between locations in ta_1. However, if ta_1 is the receiver, it can be modeled as an invariant at $i_1 \in I$, of location $l_1 \in L$, within $ta_1 \in TAV$. For the example shown in Fig. 6, gc_1, e_1 and i_1 are labeled as "event == 1".

Transformation Rule 7: Fragments
The "alt", "reg", "opt", and "break" fragments within participant $p_1 \in P$, in SDSI can be modeled as an alternative transition at $e_1 \in E$, transitioning between locations l_0 and l_1, where $l_0, l_1 \in L$, as shown in Fig. 6.

In the case of "par" fragments involving participants p_1, p_2, and p_3, where $p_1, p_2, p_3 \in P$, where the same message $m_1 \in M$, is sent from p_1 to p_2 and p_3 simultaneously, this can be modeled as a broadcast channel synchronization (denoted as $a_1 \in A$) in both sender and receiver modes, from timed automata ta_1 to both ta_2 and ta_3, where $ta_1, ta_2, ta_3 \in TAV$, simultaneously.

Furthermore, if participants p_2 and p_3 need to respond with a message to p_1, they must wait for a response message from both p_2 and p_3 before transitioning from $l_1 \rightarrow l_2$. For the example shown in Fig. 7, a_1 of ta_1 is a broadcast channel synchronization in the sender mode is labeled as "sendSignal!", while both a_1 of ta_2 and ta_3 are in the receiver mode labeled as "sendSignal?".

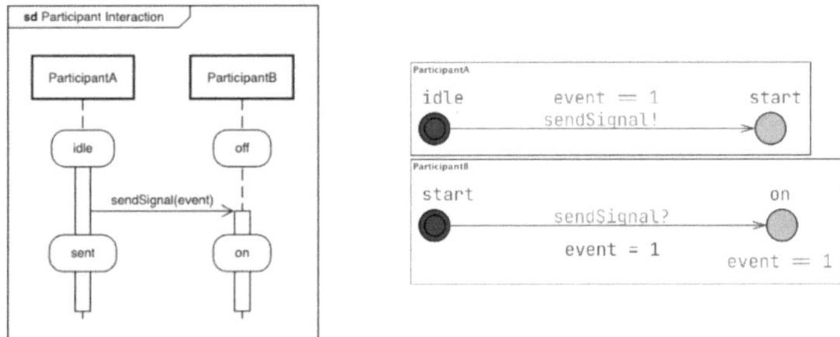

Fig. 5. Mapping a message and parameter in SDSI into timed automata.

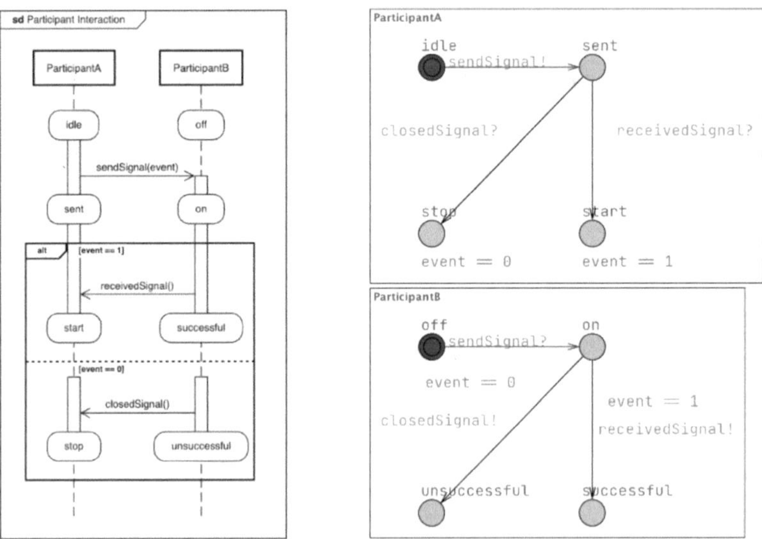

Fig. 6. Mapping the guard condition and fragments in SDSI into timed automata.

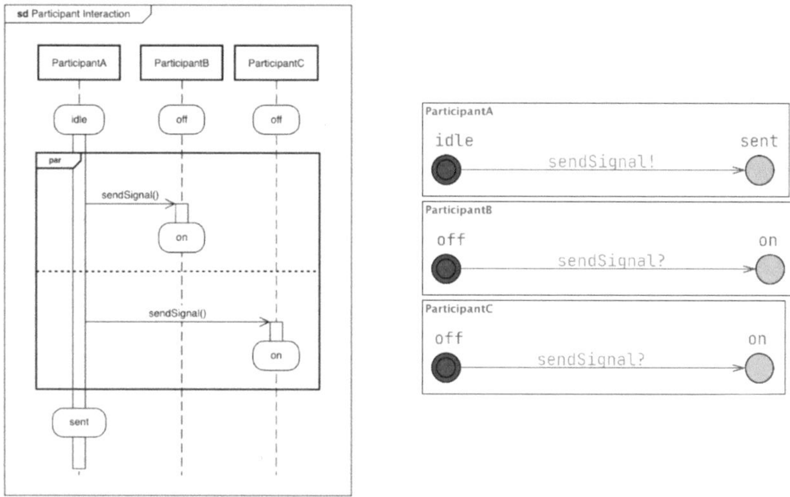

Fig. 7. Mapping the par fragment in SDSI into timed automata.

3.4 Simulate and Verify the Resulting Timed Automata

We would model timed automata using the UPPAAL tool to simulate and verify the resulting automata. These automata would be modeled using defined transformation rules. Additionally, we would define TCTL formulas to verify the timed automata for these interesting properties.

4 Case Study

In a case study, we present a case study of the fire alarm system SDSI, as shown in Fig. 8. The system includes participants, state invariants, lifelines, messages, parameters, guard conditions, and fragments. The transformed timed automata representing the fire alarm system SDSI is shown in Fig. 9. We used the UPPAAL tool to simulate the system's behavior and verify it using TCTL formulas and the verification results that demonstrate system satisfaction are shown in Fig. 10.

In this case study, we try the TCTL formula "A[] FlameSensor.Detected imply detectFire == 1" to verify the safety property. This formula asserts that in every possible system path or trace, if the flame sensor detects a fire, then the system should detect a fire, resulting in satisfaction with a green status.

The TCTL formula "A <> Controller.Notified imply MobileApp.Received" to verify the liveness property. This formula explains that in every possible system path or trace, if the controller is notified, the mobile application should receive the notification, resulting in satisfaction.

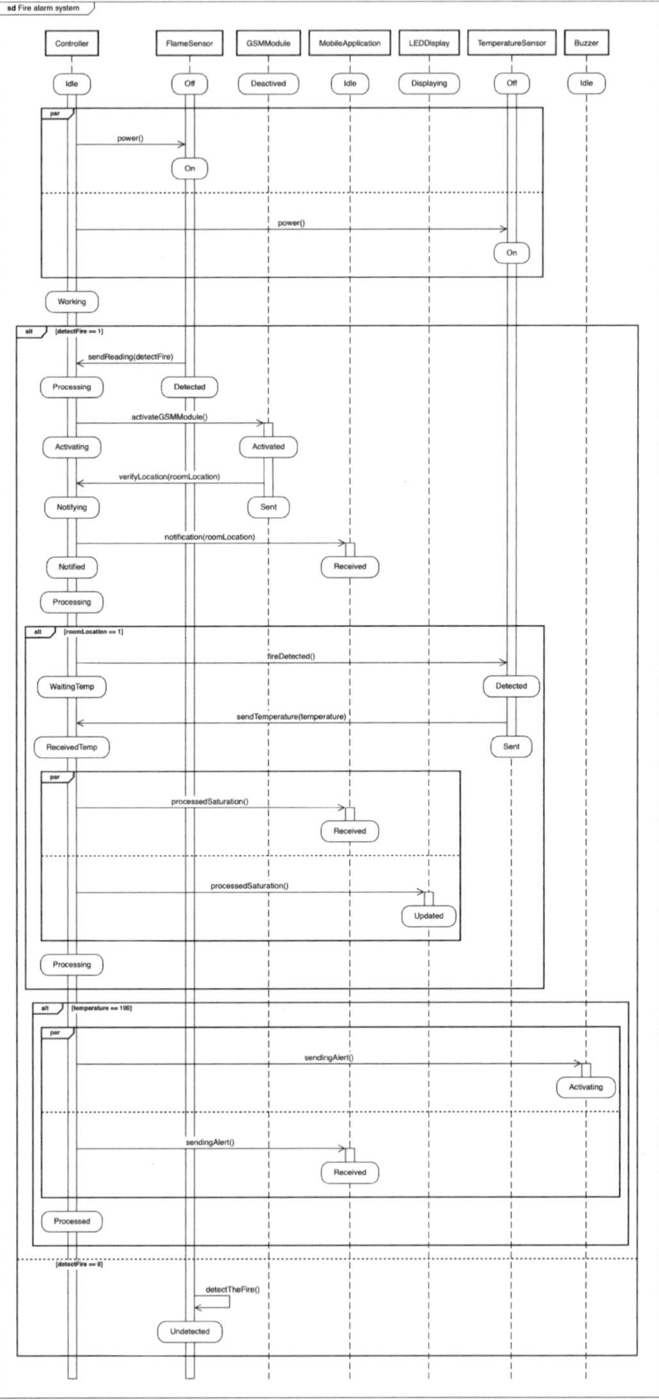

Fig. 8. The SDSI of a fire alarm system.

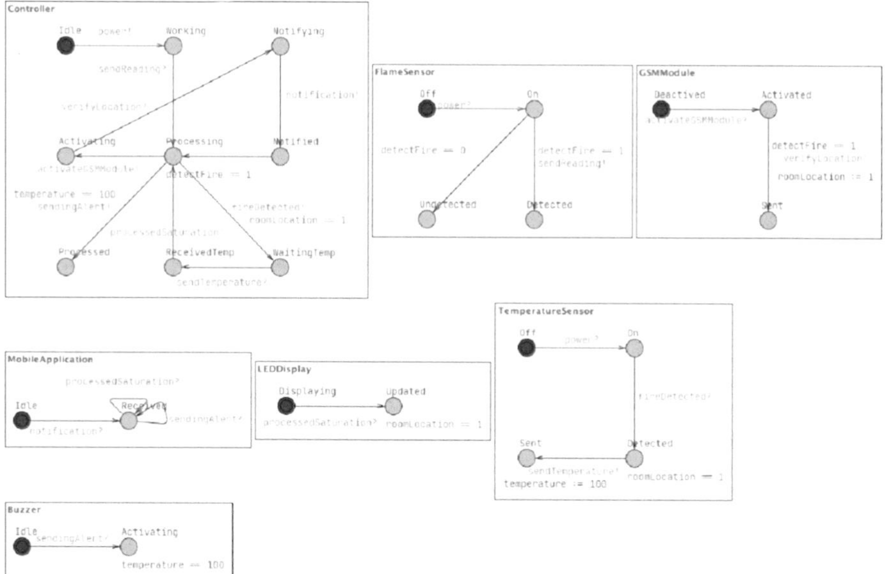

Fig. 9. The resulting timed automata from the SDSI of the fire alarm system.

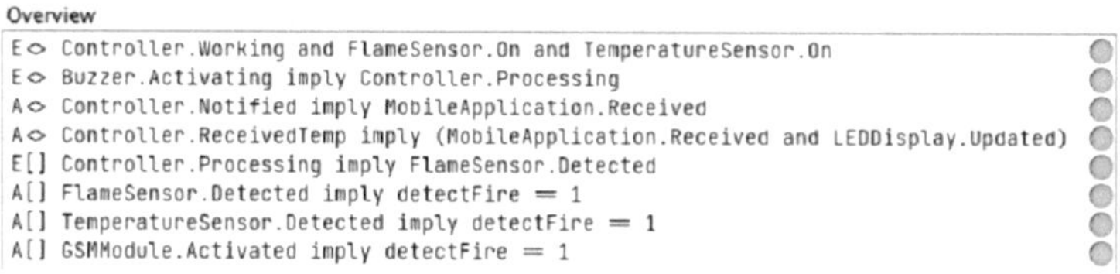

Overview

```
E<> Controller.Working and FlameSensor.On and TemperatureSensor.On
E<> Buzzer.Activating imply Controller.Processing
A<> Controller.Notified imply MobileApplication.Received
A<> Controller.ReceivedTemp imply (MobileApplication.Received and LEDDisplay.Updated)
E[] Controller.Processing imply FlameSensor.Detected
A[] FlameSensor.Detected imply detectFire == 1
A[] TemperatureSensor.Detected imply detectFire == 1
A[] GSMModule.Activated imply detectFire == 1
```

Fig. 10. The TCTL formulas and the verification results.

5 Conclusion

In this paper, we propose an approach for transforming SDSI into timed automata. We define a set of transformation rules to guide the conversion of SDSI into the corresponding timed automata. We use the UPPAAL tool to simulate and verify the resulting timed automata. We define TCTL formulas to verify these interesting properties in our case study. As part of the conversion tool development, we have already developed a software tool to convert the original SDSI into timed automata using our set of transformation rules. The resulting timed automata are correctly simulated and verified in UPPAAL tool.

References

1. ISO/IEC/IEEE 29148 Systems and software engineering – Life cycle processes – Requirement engineering (2018)
2. Russ, M., Kim, H.: Chapter 7: Modeling Ordered Interactions: Sequence Diagrams. Learning UML 2.0, pp. 108–130 (2006)
3. UML Sequence Diagrams. https://www.uml-diagrams.org/sequence-diagrams.html. Accessed 20 Oct 2023
4. OMG Unified Modeling Language version 2.5. https://www.omg.org/spec/UML/2.5/PDF. Accessed 21 Nov 2023
5. Han, D., Xing, J., Yang, Q., Wang, H., Zhang, X.: Formal sequence: extending UML sequence diagram for behavior description and formal verification. In: 2016 IEEE 40th Annual Computer Software and Applications Conference (COMPSAC), Atlanta, pp. 474–481. IEEE (2016)
6. Alur, R., Dill, D.L.: A theory of timed automata. Theor. Comput. Sci. **126**(2), 183–235 (1994)
7. Behrmann, G., David, A., Larsen, K.G.: A tutorial on UPPAAL 4.0. Department of computer science, Aalborg University (2006)
8. UPPAAL. https://uppaal.org/. Accessed 20 Oct 2023

On Students' Behavior Prediction for Library Service Quality Using Bidirectional Deep Machine Learning

Nguyen Minh Tuan[1(✉)], Phayung Meesad[2], Duong Van Hieu[3], Nguyen Ha Huy Cuong[4], and Maleerat Maliyaem[5]

[1] Department of Mathematics, King Mongkut's University of Technology North Bangkok, Bangkok, Thailand
nmtuanbm2006@gmail.com

[2] Department of Information Technology and Management, King Mongkut's University of Technology North Bangkok, Bangkok, Thailand
phayung.m@itd.kmutnb.ac.th

[3] Faculty of Engineering Technology, Tien Giang University, Ap Bac, My Tho, Tien Giang, Vietnam
duongvanhieu@tgu.edu.vn

[4] Software Development Centre, University of Danang, Le Duan, Da Nang, Vietnam
nhhcuong@sdc.udn.vn

[5] Department of Information Technology, King Mongkut's University of Technology North Bangkok, Bangkok, Thailand
maleerat.m@itd.kmutnb.ac.th

Abstract. Library service quality has been taken into account after the COVID-19 pandemic to propose appropriate conditions under the rapid change of technology circumstances. The main study in this paper is to consider the evaluation of students' sentiments to understand the role of library service quality after the pandemic outbreak and evaluate library service quality consistent with the new situation. For this study, we employ deep learning models such as Convolutional Bidirectional Long Short-Term Memory (Conv-BiLSTM) and Convolutional Bidirectional Gated Recurrent Unit (Conv-BiGRU), Attention and Transformer TFBERT model. Our findings indicate that the Conv-BiLSTM (94.59%) and Conv-BiGRU (94.33%) outperformed the others, achieving the highest accuracy for the prediction of Vietnamese students' sentiments about library service quality.

Keywords: Library service quality · Convolution model · LSTM · GRU · Attention · Transformer TfBERT

1 Introduction

Artificial intelligence (AI) development with the contribution of Natural Language Processing (NLP) achievement has set up a new age in communication

P. Meesad et al. (Eds.): IC2IT 2024, LNNS 973, pp. 55–64, 2024.
https://doi.org/10.1007/978-3-031-58561-6_6

between humans and machines. It has been applied in solving recognition problems such as ChatGPT, Chatbot, Robotics, and object detection based on AI algorithms. The algorithms banked on a large-scale dataset have been used to construe the training models such as annotated datasets, and textual datasets using state-of-the-art performances including logistic regression, support vector machine, OVO (One versus One), and OVA (One versus All) [1]. The role of sentiment has been proven that the primary transmitter of spillovers increased during the recent pandemic based on the investigation of transmitters and receivers of risk from G-7 countries [2]. To apply the model with natural language processing (NLP), word embedding is also a good choice to predict the targets using BiLSTM in the machine learning model. Embedding and Pre-train could help the BERT model to classify deceptive behaviors in the meaning of quantitative and qualitative evaluations and can propose the discriminator of the recognition of fakes [3].

Supporting machine learning is presented by the collection of datasets. A big amalgamation of the dataset including 358,743 positive reviews and 100,699 negative reviews of the Vietnamese sentiment database has been created as a good basis for predicting Vietnamese sentiment [4]. In addition, PhoBERT also can be gained for building model prediction of Vietnamese sentiment [5]. Under Twitter-based sentiment conditions, scrutinizing the correlation between necessary discretionaries supporting investors in America using Causality in quantiles [6]. Proposing a model using an aspect-level sentiment analysis model with aspect-specific context position information (CPASA) is to diminish the interference of the diversity of the meaning of words. Attention-based neural networks have been widely used to evaluate the structure of language sentiment and syntactic tendency [7].

The imbalanced datasets are prevalent in construing the prediction models. The GANs, CatGAN, and SentiGAN are usually used to synthesize the multi-data that could solve the problem [8]. The stock market and stock price are the concerns that affect investors in the holidays. By researching the investor sentiment to the fluctuation of the price in the stock market, the correlation has been shown as proof that sentiment is needed for considering investment [9]. A Topic Driven Domain Adaptation Network (TDAN) which has been built based on the attention-based neural network is the proposed model convenient for construing sentiment classification that aims for reliable classifiers [10].

Sentiment analysis has been shown by classifying comments and responses using the NaiveBayes, Logistic Regression, Gradient Boosted Trees, RandomForest, DecisionTree, NaiveBayes model based on baseline structure. Investigation of sentiment about macroeconomic expectations and consumers due to the effect of the COVID-19 pandemic, the authors have concluded that sentiment affected macroeconomics and the customers' beliefs of the government through the epidemic outbreak by comparing the customers' sentiment between the countries. Analyzing natural language by BERT, we can predict the affection of finance to human beings, and the correlation between the financial market and sentiment scores [11]. The output of analyzing text can be generated to build the Chatbot model for asking and answering between machines and humans. Compared to RNN, the LSTM model can perform the Chatbot connection more correctly than the RNN [12,13].

In this paper, the dataset UIT-VSFC [14] has been used for training to establish the prediction model, and a new dataset was collected by using an online survey to evaluate the library service quality has been extended [12,15]. We collected data through an online survey and utilized it for model training and evaluation. The main idea is to use machine learning to classify Vietnamese students' sentiments to improve service quality using deep machine learning that has been conducted and applied in predicting Vietnamese sentiments [11,16,17]. Moreover, using machine learning models to evaluate students' sentiments toward library services, particularly after the context of the COVID-19 pandemic. Now we call the first data for building model training Dataset-1 [14] (available at https://www.kaggle.com/code/duydung271/researchlearning/input) (Table 1), and the second data for classifying the Vietnamese students' assessment of service quality is Dataset-2 [15]. Dataset-2 has been collected based on the help of lecturers teaching around the universities using online Google Forms. Connection lecturers will deliver the link to students and they could take the survey in their free time. The instruction is also attached at the beginning of the form to limit the errors from the students. We have not paid for the survey and the students are not bound to respond to the questionnaire. The form is set without collecting personal information such as name, email, etc. The main content of the present paper is construed in two steps consisting of building a model from Dataset-1 and applying the built model to analyze Dataset-2.

Table 1. Short Demographical description about the Dataset-1.

text	labels
slide giáo trình đầy đủ	positive
nhiệt tình giảng dạy	positive
gần gũi với sinh viên	positive
đi học đầy đủ full điểm chuyên cần	negative
chấp nhận được	neutral
chưa áp dụng công nghệ thông tin và các thiết bị	negative
thầy giảng bài hay , có nhiều bài tập ví dụ thực tế	positive

2 Methodology

Vietnamese sentiment is considered in many aspects such as understanding customers' desires, and customers' evaluation. In this paper, Dataset-1 has been used to predict Vietnamese students' assessment of library service quality. In the past work [12], various models have been applied to translate English to

Vietnamese. We also consider the previous algorithms that work for analyzing Vietnamese sentiments in this performance. To conduct these algorithms, Dataset-1 was cleaned by removing stop-words. The remaining words were then transformed into sequences. After that, extracted words are transformed into a matrix of shapes. Moreover, some missing values are excluded and reconstructed to the same size as the test prediction. Dataset-1 shown in Table 1 consists of 11,426 instances with 2 attributes of Vietnamese sentences, and the targets are labeled positive, negative, and neutral. For Dataset-2, an online investigation has been conducted by using Google Forms. Lecturers instruct the students to access the link to evaluate the multiple choice. The students answered the questions about their behavior and library service quality assessment. They also have to explain the reasons chosen in the last comment with at least 15 words. The students' sentiment is extracted to predict Vietnamese behaviors. The investigation is based on the connection between the lecturers in the Universities around Ho Chi Minh City and Tien Giang University. The dataset attained without fee has not collected the personal information in the investigation. The full constructs are performed in the following sections:

2.1 Convolutional Bidirectional Long Short-Term Memory Networks

Long Short-Term Memory (LSTM) Networks, the specific recurrent neural network (RNN), successfully complement vanished gradient problems. The LSTM is structured by an input gate, an output gate, and a forget gate that could control the exploding gradient issues [15]. In this paper, LSTM has been combined with the layers including convolutional and bidirectional to make the new model namely Conv-BiLSTM (more detail in [12,15]). The input data has been divided into 3 sections with the ratio 60% for training, 20% for validating, and 20% testing corresponding to 6,855 sentences, 2,856 sentences, and 2,855 sentences respectively. LSTM is established with 6 layers consisting of the input layer, convolution 1D, Maxpooling 1D, Spatial Dropout 1D, Convolutional Bidirectional LSTM, and Dense. The total parameters of the model are 106,720 parameters as shown in Table 2. The input and output shapes chosen consistently according to the max length and min length of the data in this model are (None, 250, 64) and (None, 32) respectively.

2.2 Convolutional Bidirectional Gated Recurrent Unit

Similar to LSTM, a Gated Recurrent Unit (GRU) represents the form of recurrent neural networks with an existing input gate, and forget gate. For this model, Dataset-1 is used for building prediction models, consisting of 11,426 instances. The model has been built using data cleaned and split into 3 sections of the ratio 60% of the training set, 20% validation set, and 20% testing set corresponding to 6,855 sentences, 2,856 sentences, and 2,855 sentences respectively. GRU is established with 6 layers consisting of the input layer, convolution 1D, Maxpooling 1D, Spatial Dropout 1D, Convolutional Bidirectional GRU, and Dense. The total parameters of the model are 90,592 parameters performed in Table 3. Regarding the data features mentioned above, the input and output shapes chosen consistently in this model are (None, 250,64) and (None, 32), respectively.

2.3 Attention

Structured by the pre-train convolutional neural network, Attention is a kind of recurrent neural network and has been built in the form of an encoder-decoder system. The convenient function of Attention is to solve with a huge number of inputs and outputs. For this model, Dataset-1 is processed the same way as LSTM, including 11,426 sentences. The input data has also been cleaned to omit punctuation, special characters, stop words, etc., by the Tokenizer function, and divided into 3 sections ratio 60% of the training set, 20% validation set, and 20% testing set corresponding to 6,855 sentences, 2,856 sentences, and 2,855 sentences respectively. The Attention model is built with six layers. These include an input layer, an embedding layer, an Attention layer, a global_average_pooling1d layer, and two Dense layers. The total parameters of the model are 64,760 parameters delineated in Table 2. Based on the features of the data, the input and output shapes that have been chosen suitably with this model are 2 dimensions (None, None), and (None, 10), respectively.

2.4 Transformer TFBERT

The pre-train BERT model could be created with a fine-tuned function to enhance the quality of the prediction. BERT has been confirmed effective in analyzing NLP. BERT is built to leverage the understanding of the context by a sequence of blocks of encoders. The input sentences are tokenized and transformed into embedding vectors where BERT could produce the output vectors. In this work, the Transformer model is built with 4 layers embracing Input, TFBERTModel, SlicingOpLambda, and Dense layer. The total parameters of this model consist of 109,082,113 parameters marked out in Table 2. The input shape for this model is (None, 192) and the output shape is (None, 1). Regarding the features of the models mentioned above, the input and output shapes chosen consistently in this model are (None, 250, 64) and (None, 32), respectively.

3 Experiment Results

Using Dataset-1, the models were built after cleaning to establish new data including 11,426 Vietnamese sentences attached to 3-labeled targets (positive, negative, and neutral). The meaning of these sentences are Vietnamese sentiment evaluation including good (tốt, chuẩn, tuyệt miễn bàn, ...), neutral (bình thường, không ý kiến, không quan tâm, ...), and negative (không tốt, tệ quá, chưa tốt, cònthiếu thốn, ...). For the evaluation of Dataset-2 [15], an online survey has been conducted, and the dataset is constructed with multiple choice consisting of 29 attributes and explanations filled with at least 15 words (students are required to fill) to show the reasons that they chose target options. The library service was assessed using three options: good (positive), normal (neutral), and bad (negative). Among the respondents, 1,257 students marked 'good', 826 marked 'normal' (with 92 missing labels manually filled), and 203 marked 'bad'. The survey was collected from 3,450 participants, most of the students were freshmen and sophomores. The reason for these cases is based on the structure of the first and the second year, students are studied in general coursework and they could come to the library frequently. In the third and fourth years, students usually focus on internships or practically learn in the company. The missing values are excluded in this work and have been rescaled consisting of 2,258 students' comments. We conducted an extra convey for the complement data and reconstructed input data to get the same size (2,286 inputs) compared to the validation prediction of the model for Dataset-1. The data prediction has been shown in Table 4.

3.1 Model Evaluation

In this section, we illustrate the evaluation results collected using the deep machine learning models consisting of Conv-BiLSTM, Conv-BiGRU, Attention, and Transformer TFBERT. The metrics used in this model are accuracy and loss performed in Eq. (1), and Eq. (2). Accuracy is calculated by the total number of right-predicted results in the total number of predicting contexts. The accuracy formula is shown in Eq. (1).

$$Accuracy = \frac{TP + TN}{TP + TN + FP + FN} \tag{1}$$

where TP be abbreviation of positive samples, FP be abbreviation of mistake positive samples, TN be abbreviation of negative samples, and FN be abbreviation of mistake negative samples. The models are for calculating the errors and the odds of the analyzed algorithms whose results gained probability is in the interval [0,1]. The basic classified categorical cross-entropy loss prediction is shown in Eq. (2).

$$L_{cce} = \frac{-1}{M} \sum_{k=1}^{K} \sum_{m=1}^{M} y_m^K log(h_\theta(x_m, k)) \tag{2}$$

where M stands for the number of training samples, K stands for the number of classes, y_k^m stands for the target labels for the training examples m for class k, x denotes input for training samples m, and h_θ stands for recurrent neural network weights.

3.2 Convolutional Bidirectional Long Short-Term Memory Networks

We used Dataset-1 to train the above model for 50th epochs. The training process relied on metric accuracy for numerical prediction (Fig. 1). The output value of this model gained 0.9459 for accuracy, 0.1210 for loss, and 0.7837, and 0.8381 for validation sets respectively (shown in Table 3). The time for processing is 51 s. The performance of the prediction for this model is depicted in Fig. 1. The model is transformed to predict Dataset-2 and achieves the output with 1,254 students (good), 995 students (neutral), and 37 students (negative) (shown in Table 4) (Fig. 2).

3.3 Convolutional Bidirectional Gated Recurrent Unit

Applying the model built and Dataset-1 scaled, considering the 50th epoch, the data prediction has been reached including 0.1197 for loss, 0.9433 for accuracy, 0.7691 for validation loss, and 0.8258 for validation accuracy (Table 3). The time running per epoch is 27 s. This model is transformed to predict Dataset-2 and successfully performed with the results consisting of 1,168 students (good), 1,059 students (neutral), and 59 students (negative) (Table 4, Fig. 3).

3.4 Attention

Using Dataset-2 and the model built above, the process is trained for 50th epochs to get the numerical outputs (Fig. 3). The metrics applied in this section attained 0.8066 for the loss, 0.5898 for accuracy, 0.7809 for validation loss, and 0.5978 for validation accuracy. Time costing for an epoch is 16 s (Table 3). The prediction for Dataset-2 performed in Table 4 consists of 1,257 students (positive), 734 students (neutral), and 203 students (negative).

3.5 Transformer TFBERT

Dataset-2 is applied in this model after being scaled consistent with the test set of Dataset-1. The model is trained for 50th epochs concerning 13 s per epoch (Fig. 4). The output of prediction consists of a loss of 0.1210, accuracy of 0.4706, validation loss of 0.1310, and validation accuracy of 0.4475 (Table 3). The Dataset-2 prediction for this model is performed in Table 4 including 17 students (positive), 2,269 students (neutral), and 0 students (negative).

Table 2. Model setting.

Models	Input, Output shape	Layers	Parameters
Conv-LSTM	(None, 250, 64), (None, 32)	6	106,720
Conv-GRU	(None, 250, 64), (None, 32)	6	90,592
Attention	[(None, None), (None, None)], (None, 10)	6	64,760
TransformerTFBERT	(None, 192), (None, 1)	4	109,082,113

4 Limitation and Discussion

The role of the convolution layer has been effectively shown in increasing the accuracy of predicting present Vietnamese students' sentiments. The convo-

Table 3. Numerical output of Dataset-1.

Models	Loss	Accuracy	Val_loss	Val_accuracy	Time per Epoch (s)
ConvBiLSTM	0.1210	0.9459	0.7837	0.8381	51
ConvBiGRU	0.1197	0.9433	0.7691	0.8258	27
Attention	0.8066	0.5898	0.7809	0.5978	16
TransformerTFBERT	0.1210	0.4706	0.1310	0.4475	13

Table 4. Model predictions for Dataset-2.

Models	Students' assessment	Conv-LSTM prediction	Conv-GRU prediction	Attention prediction	TFBERT prediction
Good	1,257	1,254	1,168	152	17
Neutral	826	995	1,059	2,134	2,269
Negative	203	37	59	0	0

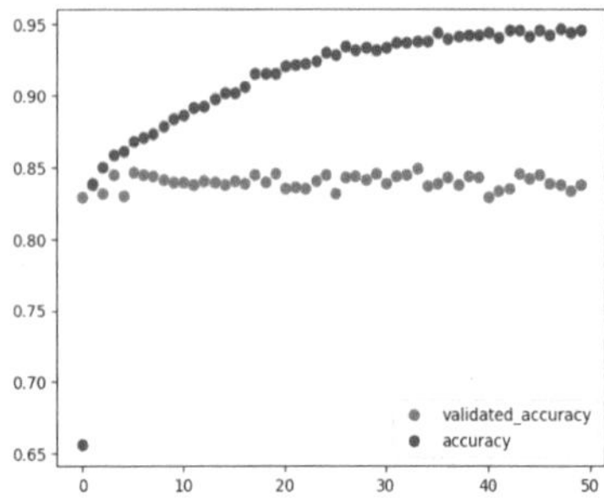

Fig. 1. LSTM Prediction.

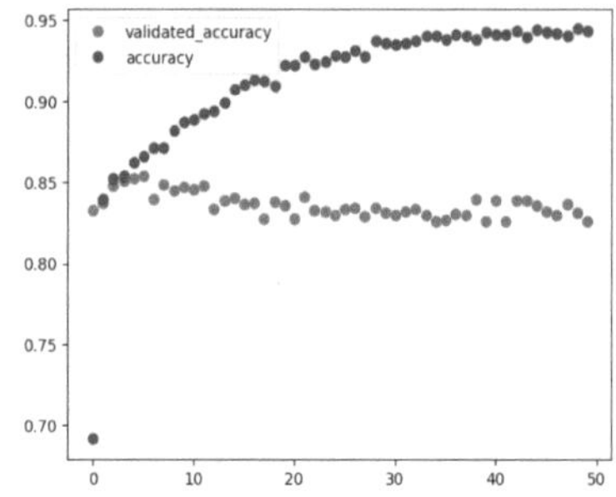

Fig. 2. GRU prediction.

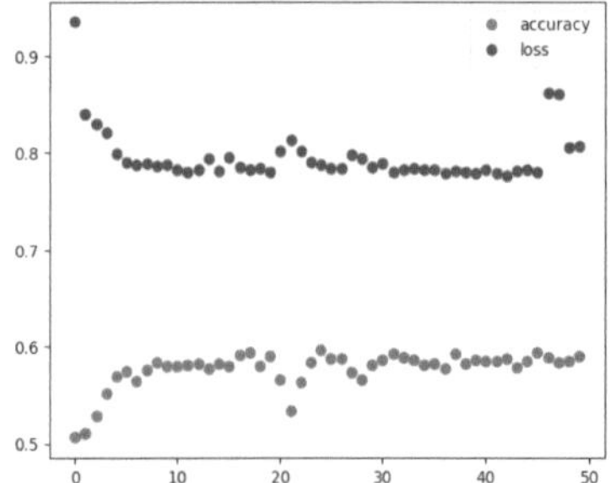

Fig. 3. Attention Prediction.

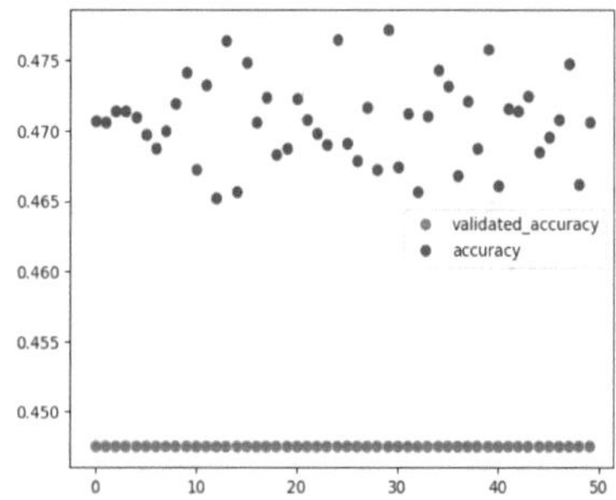

Fig. 4. TransformerTfBERT prediction.

lution layer is construed with arrays of input kernels and operates between the filters and input data through weights. These features make the difference to other layers and the success of the prediction including the combination of Bidirectional LSTM and Bidirectional GRU. The bias situation could happen when most of the students only choose the 'good' target or 'neutral' targets. To fix the bias of the data, the instruction is given at the beginning of the questionnaire that could avoid the bias situation. In the present case of data bias, we have made an instruction and extra survey to get the same size to be consistent with the validation and test samples of the models. The positive comments focused on the cleanness (sạch sẽ, thoáng mát, ngăn nắp,...), staff behavior in library (thái độ phục vụ nhân viên), computer for searching (hỗ trợ máy tính cho việc tìm kiếm tài liệu), and the infrastructure with technology (hỗ trợ thêm công nghệ) could be applied to improve the library service quality.

5 Conclusion and Future Work

The paper has successfully evaluated Vietnamese sentiment under the collection of students' assessments about service quality. Using deep machine learning, some models including Conv-BiLSTM, Conv-BiGRU, Attention, and Transformer have been built to evaluate Vietnamese students' sentiments about library service quality. The findings indicate that the Conv-BiLSTM model outperformed the others, achieving the highest accuracy. The employed models effectively classify students' sentiments, providing valuable insights into their perceptions of library services. The analyzed results of the students' assessment are suitable references that the official management of the university could consider to improve the library service quality and general service quality in the age of development of technology and the run of educational achievement.

References

1. Duong, H.-T., Nguyen-Thi, T.-A., Hoang, V.T.: Vietnamese sentiment analysis under limited training data based on deep neural networks. Complexity **2022**, 1–14 (2022). https://doi.org/10.1155/2022/3188449
2. Aharon, D.Y., et al.: Related COVID-19 media sentiment and the yield curve of G-7 economies. North Am. J. Econ. Financ. **61** (2022). https://doi.org/10.1016/j.najef.2022.101678
3. Catelli, R., et al.: Deceptive reviews and sentiment polarity: effective link by exploiting BERT. Expert Syst. Appl. **209** (2022). https://doi.org/10.1016/j.eswa.2022.118290
4. Nguyen, C.V., et al.: Learning for amalgamation: a multi-source transfer learning framework for sentiment classification. Inf. Sci. **1–14**, 590 (2022). https://doi.org/10.1016/j.ins.2021.12.059
5. Dang, R., et al.: Sentiment analysis for Vietnamese - based hybrid deep learning models. Comput. Sci. Math. **2023** (2022). https://doi.org/10.20944/preprints202306.1318.v1

6. Zeitun, R., et al.: The impact of Twitter-based sentiment on US sectoral returns. North Am. J. Econ. Financ. **64** (2023). https://doi.org/10.1016/j.najef.2022.101847
7. Huang, B., et al.: CRF-GCN: an effective syntactic dependency model for aspect-level sentiment analysis. Knowl.-Based Syst. **260**, 110125 (2023). https://doi.org/10.1016/j.knosys.2022.110125
8. Imran, A.S., et al.: The impact of synthetic text generation for sentiment analysis using GAN based models. Egypt. Inform. J. **23**(3), 547–557 (2023). https://doi.org/10.1016/j.eij.2022.05.006
9. Liu, Q., Huang, M., Zhao, L., Lee, W.-S.: The dispositional effects of holidays on investor sentiment: therapeutic and hygienic. J. Innov. Knowl. **8**(2), 100358 (2023). https://doi.org/10.1016/j.jik.2023.100358
10. Zhu, Y., et al.: Topic driven adaptive network for cross-domain sentiment classification. Inf. Process. Manag. **60**(2), (2023). https://doi.org/10.1016/j.ipm.2022.103230
11. Tuan, N.M., Meesad, P.: A study of predicting the sincerity of a question asked using machine learning. In: 5th International Conference on Natural Language Processing and Information Retrieval (NLPIR), pp. 129–134 (2021). https://doi.org/10.1145/3508230.3508258
12. Tuan, N.M., Meesad, P., Nguyen, H.H.C.: English-Vietnamese machine translation using deep learning for chatbot applications. SN Comput. Sci. **5** (2024). https://doi.org/10.1007/s42979-023-02339-2
13. Meesad, P.: Thai fake news detection based on information retrieval, natural language processing and machine learning. SN Comput. Sci. **2**(6), 425 (2023). https://doi.org/10.1007/s42979-021-00775-6
14. Nguyen, K.V., et al.: UIT-VSFC: Vietnamese students' feedback corpus for sentiment analysis. In: 10th International Conference on Knowledge and Systems Engineering (KSE), pp. 19–24 (2018). https://doi.org/10.1109/KSE.2018.8573337
15. Tuan, N.M., et al.: New data about library service quality and convolution prediction. CTU J. Innov. Sustain. Dev. **14** (2023). https://doi.org/10.22144/ctujoisd.2023.032
16. Minh, T.N., Meesad, P., Nguyen Ha, H.C.: English-Vietnamese machine translation using deep learning. In: Meesad, P., Sodsee, S., Jitsakul, W., Tangwannawit, S. (eds.) IC2IT 2021. LNNS, vol. 251, pp. 99–107. Springer, Cham (2021). https://doi.org/10.1007/978-3-030-79757-7_10
17. Tuan, N.M.: Machine learning performance on predicting banking term deposit. In: Proceedings of the 24th International Conference on Enterprise Information Systems, pp. 267–272 (2022). https://doi.org/10.5220/0011096600003179

Formal Verification of Vessel Scheduling Using Probabilistic Timed Automata

Ratchanok Thianpunyathanakul(✉) and Wiwat Vatanawood

Department of Computer Engineering, Faculty of Engineering, Chulalongkorn University, Bangkok, Thailand
6470263821@student.chula.ac.th, wiwat@chula.ac.th

Abstract. The effective management of time delays in vessel scheduling within the vessel transportation industry poses a complex challenge, as potential delays can lead to shifts in planned routes and increased costs. This paper addresses this issue by proposing a formal probabilistic model for vessel scheduling that considers time constraints and accommodates relevant risk factors contributing to delays. Utilizing Probabilistic Timed Automata (PTAs), our model captures distributed random delays from key risk factors such as weather conditions, mechanical engine readiness, and port congestion in the target vessel scheduling. We then translate PTAs model representation of the target vessel scheduling into PRISM code. Subsequently, the PRISM model checker is employed to simulate and verify the vessel scheduling model's compliance with specified behavioral properties expressed in Probabilistic Computation Tree Logic (PCTL). This approach facilitates a comprehensive analysis of time-related aspects, aiding in both planning and mitigation of associated risk impacts.

Keywords: Probabilistic Timed Automata · Probabilistic Computation Tree Logic · Discrete-Time Markov Chain · Vessel Scheduling

1 Introduction

At present, the management of vessel transportation systems [1] and scheduling encounters intricate challenges in addressing the uncertainties tied to randomly distributed delays and deviations from planned routes in vessel operations. These deviations from established schedules pose significant obstacles, resulting in increased costs stemming from delayed arrivals. The fundamental factors contributing to these challenges are the uncertainties and variations in the temporal aspects of vessel operations, crucial elements in the strategic planning and efficient management of vessel schedules. Several research studies focused on the management of vessel transportation systems, employing simulation methodologies in some cases, while others chose to formalize vessel scheduling into models and conduct the relevant optimization algorithms or using genetic algorithm (GA) [2] which is an optimization and search algorithm inspired by the principles of natural evolution and genetic selection. Aimed at vessel scheduling, employing GA [2] to generate and evaluate a series of potential solutions. The algorithm helps in finding

P. Meesad et al. (Eds.): IC2IT 2024, LNNS 973, pp. 65–72, 2024.
https://doi.org/10.1007/978-3-031-58561-6_7

the most suitable positions for vessels, aligning them optimally with the movements of vessel shipment and other feeder vessels within the port. Generally, GA operates by continuously comparing and improving multiple solutions, evolving towards the best possible outcome, GA efficiently navigates through a multitude of possible solutions, identifying the most effective strategies for managing vessel traffic and port operations to search for the effective vessel scheduling and planning.

However, in addition to the studies previously mentioned, our focus is on addressing these challenges through a formal verification approach. In this paper, we propose an alternative method to formalize the target vessel scheduling, incorporating time constraints and the pertinent risk factors contributing to time delays namely, weather conditions, mechanical engine readiness, and port congestion. We employ Probabilistic Timed Automata (PTAs) [3, 4] to represent the target vessel scheduling. The PTAs representation is then translated into PRISM code, utilizing the PRISM programming language to articulate the structure and behavior of the PTAs that encapsulates our vessel scheduling. Subsequently, we simulate the resulting PRISM code using the PRISM model checker tool. Our case study focuses on the PTAs of vessel scheduling from Thailand port to Singapore port, where three identified risk factors associated with time delay weather conditions, mechanical engine readiness, and port congestion at the destination port are considered. Compliance properties to be verified are expressed using Probabilistic Computation Tree Logic (PCTL) formula.

This paper is organized as follows. Section 1 is the introduction. Section 2 introduces technical background. Our formal verification method is described in Sect. 3. And Sect. 4 is our conclusion.

2 Background

2.1 Probabilistic Timed Automata

PTAs [3] is a collaborative model that combines the structure of timed automata and probabilistic modeling, specifically incorporating timed automata with Markov decision process (MDP). The integration of timed automata aids in accurately modeling and analyzing vessel operations during significant time intervals. On the other hand, MDP is a mathematical model used for decision-making in situations with uncertainties and statistical relationships. In the modeling of vessel scheduling, PTAs model can be transformed into MDP to plan decision-making for route selection and management to induce delays in a suitable format. The actions and states can be specified based on time, allowing the simulation of operations with dynamically changing environments. Additionally, the consideration of probabilities in state changes enables the modeling and analysis of systems with time variations and random variable actions. Probabilistic timed automata (PTAs) is a timed automata with the addition of probabilities in state transitions. A sample PTAs model of message sending behavior is shown in Fig. 1. A message is supposed to be sent and received at the destination with time constraint. A variable clock y is counting to indicate the time out state when $y \geq 20$. The possibility of the complete message sending is 0.9, while the loss of the message sending is 0.1. The retry of the message sending occurred while the global clock y is counting, as example are shown in Fig. 1.

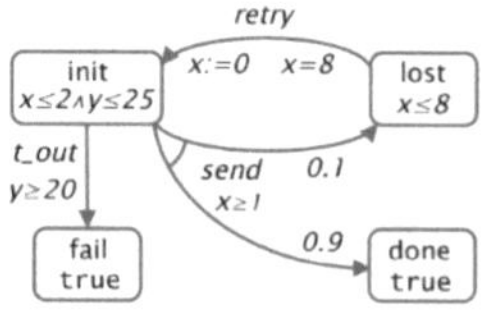

Fig. 1. A sample PTAs model of message sending behavior [3].

2.2 PRISM Model Checker

PRISM [5] is a probabilistic model checking tool used to analyze and verify models like PTAs. It plays a critical role in this paper, enabling the simulation and verification of the vessel scheduling model. PRISM's ability to handle complex probabilistic models makes it an ideal tool for assessing the intricacies of vessel scheduling, where numerous variables and uncertainties need to be considered. Therefore, it is a powerful tool for analyzing and simulating systems modeled with mathematical precision, particularly those involving decision-making and probabilistic behavior. It supports a range of mathematical models, such as control systems, communication systems, computer systems, and others. For this paper, PRISM version 4.7 [5] will be utilized as the model checking tool for probabilistic models. PRISM is renowned for creating and analyzing models of probabilistic behaviors, and it is commonly used in various application domains. The official PRISM manual [5] provides detailed information on using PRISM, including the PRISM language for specifying mathematical models.

3 Our Simulation Approach

In our alternative approach, we formalize the target vessel scheduling into a Probabilistic Timed Automata (PTAs) model. The PTAs model is then translated into PRISM code and simulated. The vessel scheduling model, represented in PRISM code, undergoes iterative simulations to ensure that the probability values align with historical data. The PRISM code representing the vessel scheduling is then verified using Probabilistic Computation Tree Logic (PCTL) [5] formula. The PTAs model represents vessel scheduling based on three key factors, and the PRISM Model Checker is used to test and verify whether the PTAs model's probabilities align with the code in PRISM to enhance the overall effectiveness and stability of vessel transportation scheduling systems in addressing the challenges posed by uncertainties and dynamic factors. This approach involves several steps, as below.

3.1 PTA Models of Vessel Scheduling

To model a Probabilistic Timed Automata (PTAs) of vessel scheduling, we establish the initial state as the vessel's position at the original port, and the final state corresponds to the destination port. The state transitions between the initial and final states incorporate probabilistic considerations for various risk factors. In our PTAs model, three specific risk factors are considered. Moreover, a global clock variable named "*recdelay*" is defined to collect the total time delay from the initial to final state.

Firstly, we address time delays that may arise due to weather conditions. In Fig. 2, the transition from state 1 to state 2 represents the expectation of normal weather conditions with a corresponding probability value. Conversely, the transition from state 1 to state 3 signifies the anticipation of adverse weather conditions, also associated with a distinct probability value. In our PTA, the distributed random delays are defined as the iterative path of state transitions from state 3 back to state 1. Secondly, for the time delay due to mechanical engine readiness, the transition from state 4 to state 5 represents the expectation of good conditions of the mechanical engine, associated with a defined probability value. Conversely, the transition from state 4 to state 6 indicates the anticipation of bad mechanical engine conditions with a distinct probability value. Similar to the weather-related delays, the distributed random delays in this context are defined as the iterative path of state transitions from state 6 back to state 4 as well.

We employ the same state transition path patterns used for weather conditions to model port congestion conditions. Similarly, in the context of port congestion, the transition from one state to another reflects different congestion scenarios, each associated with distinct probability values. This consistency in approach facilitates a comprehensive representation of the probabilistic aspects associated with both weather conditions and port congestion within the vessel scheduling model.

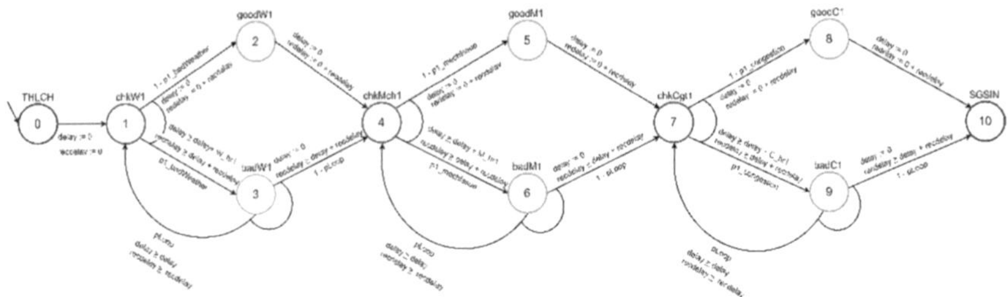

Fig. 2. The PTA model of the vessel scheduling of Thailand port to Singapore port.

In a PTA model shall be considered the probability values for distributed random delays to implement in PRISM code, the formal verification model explores all possible transitions of the DTMC model [6, 7] apply to compute PCTL on PRISM model checker tool of vessel transport scheduling in which the better-than-worst-case timing constraint [6]. We provide the flexibility for the end user to input the initial values of the probability values associated with each state transitions outlined in Sect. 3.1. For example, the transition from state 1 to state 2 requires a probability value denoting the expectation of the normal weather condition, referred as "*p1_badWeather*". Similarly, the state transition from state 3 back to state 1 needs a probability value for the iterative path, referred as "*pLoop*". As shown in Fig. 4(a), and also simulation PTAs model onto PRISM through PCTL formulas for Properties of PRISM, as shown in Fig. 4(b), correspond to the likelihood of encountering specific conditions that can delay a vessel's journey. The initial probability values of "*p1_badWeather*", "*p1_mechIssue*", "*p1_congestion*", and "*pLoop*" are filled up with 0.8, 0.1, 0.1, 0.5 respectively. These initial probability values are chosen based on the historical data and expert recommendation. For example, the initial probability value for the expectation of bad weather, denoted as "*p1_badWeather*

= *0.8*", signifies that the time delay caused by the adverse weather condition is higher. This suggests that the weather condition is anticipated to be bad with an 80% likelihood. While "*p1_mechIssue* = *0.1*" means that a lower likelihood of time delays due to mechanical engine issues, indicating an anticipated occurrence with a 10% probability. Similarly, for "*p1_congestion* = *0.1*", the lower likelihood of time delays is associated with port congestion, indicating an anticipated occurrence with a 10% probability as well.

```
dtmc
const int maxDays = 11; // Maximum number of days specified
const int maxHours = maxDays*24; // Maximum number of days specified
const int numPorts = 4;// Maximum number of Ports terminal specified
const int maxDelay = maxHours; // Maximum delay at a port
formula allPorts = 10*(numPorts)+1;
const int locations;// state ports visited - Maximum at a port sailing
const double pLoop;// prob. value for delay recursive once found bad factors
const double p1_badWeather;
const double p1_mechIssue;
const double p1_congestion;
module VesselSailing
loc : [0..allPorts] init 0;  delay : [0..maxDelay] init 0;  recdelay : [0..maxDelay] init 0;
[sail_port_THLCH] (loc=0) -> 1: (loc'=1) & (delay'=0)& (recdelay'=0);
[chkWeather_THLCH] (loc=1)  -> 1-p1_badWeather : (loc'=2) &(delay'=min(0,maxDelay))& (recdelay'=min(0+recdelay,maxDelay))
+ p1_badWeather : (loc'=3)& (delay'=min(6,maxDelay));
[goodWeather_THLCH] (loc=2) -> 1: (loc'=4) &(delay'=min(0,maxDelay))& (recdelay'=min(0+recdelay,maxDelay));
[badWeather_THLCH] (loc=3) -> 1-pLoop: (loc'=4) &(delay'=min(delay,maxDelay))& (recdelay'=min(delay+recdelay,maxDelay))
+pLoop: (loc'=1) &(delay'=min(delay,maxDelay))& (recdelay'=min(delay+recdelay,maxDelay));
[chkMechIssue_THLCH] (loc=4) -> 1-p1_mechIssue : (loc'=5) &(delay'=min(0,maxDelay))& (recdelay'=min(0+recdelay,maxDelay))
+ p1_mechIssue : (loc'=6) & (delay'=min(1,maxDelay));
[goodMechIssue_THLCH] (loc=5) -> 1: (loc'=7) &(delay'=min(0,maxDelay))& (recdelay'=min(0+recdelay,maxDelay));
[badMechIssue_THLCH] (loc=6) -> 1-pLoop: (loc'=7) &(delay'=min(delay,maxDelay))& (recdelay'=min(delay+recdelay,maxDelay))
+pLoop: (loc'=4) &(delay'=min(delay,maxDelay))& (recdelay'=min(delay+recdelay,maxDelay));
[chkCongestion_THLCH] (loc=7) ->1-p1_congestion : (loc'=8) &(delay'=min(0,maxDelay))& (recdelay'=min(0+recdelay,maxDelay))
+ p1_congestion : (loc'=9) & (delay'=min(8,maxDelay));
[goodCongestion_THLCH] (loc=8) -> 1: (loc'=10) &(delay'=min(0,maxDelay))& (recdelay'=min(0+recdelay,maxDelay));
[badCongestion_THLCH] (loc=9) -> 1-pLoop: (loc'=10) &(delay'=min(delay,maxDelay))& (recdelay'=min(delay+recdelay,maxDelay))
+pLoop: (loc'=7) &(delay'=min(delay,maxDelay))& (recdelay'=min(delay+recdelay,maxDelay));
[arrive] (loc=10) -> (loc'=11);
endmodule
```

Fig. 3. A Sample of PRISM code.

Model Constants

Name	Type	Value
locations	int	11
pLoop	double	0.5
p1_badWeather	double	0.8
p1_mechIssue	double	0.1
p1_congestion	double	0.1

(a)

Properties

P=? [F loc=locations&(recdelay>0)]

P=? [F loc=locations&(recdelay=0)]

(b)

Fig. 4. Initializing the probability values associated with each state transitions in PTAs for PRISM code (a), Properties contain PCTL formula for PRISM code (b).

3.2 Translate PTAs into PRISM Code

In this step, the PTAs model of vessel scheduling is successfully translated into PRISM code. Every element of the PTAs encompassing state, transition, clock variable, guard condition and probability value condition, is translated into statements within PRISM programming language. The resulting PRISM code is structured into two primary parts: first, the definition section, which includes variables, constants and their initial values; second, the section detailing the connections between one particular state and its subsequent states along with the guard conditions and the corresponding actions to be executed. A sample of PRISM code is shown in Fig. 3.

3.3 Simulate and Verify the Behavioral Properties

The resulting PRISM code translated from the PTAs model is now representing the vessel scheduling from Thailand port to Singapore port. Three key risk factors are concerned as weather conditions, mechanical engine readiness, and port congestion. We now employ the PRISM model checker tool to do the simulation and verification of this vessel scheduling model. As we mentioned earlier, PCTL formula can be used to articulate specific conditions and performance metrics that the PTAs vessel scheduling model should meet or predict. Moreover, the PCTL formula includes traditional temporal logic operators such as path quantifiers (A for "*for all*", E for "*there exists*") and temporal operators (X for "*next*", F for "*eventually*", G for "*globally*"), and Boolean connectives ($\bigwedge$, $\bigvee$, $\neg$). PCTL formula also introduces probabilistic operators, including $Pr \sim p(\emptyset)$ to denote "*the probability of satisfying ∅ is at least p*" so that the simulation with discretionary distributed random delays can be performed to reflect the real-world data [1, 2] from uncertainties in vessel transport scheduling constraints such as weather conditions, mechanical engine readiness, and port congestion.

We begin the verification process by entering the PCTL formula describing how specific conditions and outcomes can be quantified and analyzed within the given PRISM code of the vessel scheduling model as shown in Fig. 4(b). In this case, users can specify formulas such as "*P = ? [F loc = locations & (recdelay > 0)]*" and "*P = ? [F loc = locations & (recdelay = 0)]*", where "*P*" is the probability value to be calculated, "*loc = 1*" indicates Thailand port/"*loc = 10*" indicates Singapore port, and "*recdelay*" indicates the total time delay.

The formula "*P = ? [F loc = locations & (recdelay > 0)]*" indicates the calculation of the probability of the vessel reaching the location "*loc*" with the total time delay, accumulated in clock variable named "*recdelay*", exceeding zero hours. If the resulting probability *P = 0.8*, it signifies that there is an 80% probability of the vessel reaching the specified location "loc" with a time delay.

The formula "*P = ? [F loc = locations & (recdelay = 0)]*" indicates the calculation of the probability of the vessel reaching the location "*loc*" as planned with no time delay when the clock variable named "*recdelay*" is equal to zero. If the resulting probability "*P = 0.2*", it signifies a 20% probability of the vessel reaching the specified location "*loc*" on time.

Together, these PCTL formulas provide a comprehensive view of the vessel scheduling simulation, demonstrating the model's capability to predict and analyze various scenarios. They also underscore the importance of probabilistic modeling in handling uncertainties in complex logistical operations like vessel scheduling. Furthermore, the PRISM model checker tool generates a trend graph illustrating the probability value concerning the location "*loc*". As shown in Fig. 5, our focus is on the location "*loc = 10*", which represents the destination at the Singapore port. The simulation and verification results, in accordance with the provided PCTL formulas mentioned earlier, reveal that under the current conditions, there is a likelihood of the vessel scheduling being delayed.

Specifically, the graph indicates a probability of only 16% for the vessel to reach the destination at the Singapore port on time.

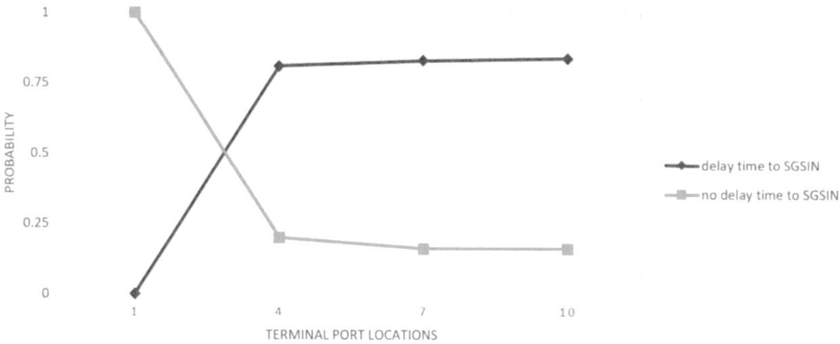

Fig. 5. Trend graph of probabilities that are shown delay time and no delay time.

4 Conclusion

In this paper, we propose an alternative approach to formalize the target vessel scheduling from the original port to the destination port into a Probabilistic Timed Automata (PTAs) model. This PTAs comprehensively captures distributed random delays stemming from critical risk factors, such as weather conditions, mechanical engine readiness, and port congestion at the destination port. The PTAs model is then translated into PRISM code and subjected to simulation using historical data from the vessel transportation database. The vessel scheduling model, represented in PRISM code, undergoes iterative simulations to ensure that the probability values align with historical data. Subsequently, the PRISM model becomes proficient in predicting the random delay time of the vessel scheduling. The model's accuracy is then verified using Probabilistic Computation Tree Logic (PCTL) formulas.

References

1. RLI: LAEM CHABANG - INDIA SERVICE. https://www.rclgroup.com/pdf/SvcPtrn/20230314-RLI.PDF. Accessed 01 Nov 2023
2. Belfkih, A., Diarrassouba, I., Yassine, A., Joncour, C.: A model for optimal feeder vessel management in container terminals. In: 4th International Conference on Logistics Operations Management (GOL), pp. 1–6. IEEE, Le Havre (2018)
3. Model Checking for Probabilistic Timed Automata. https://www.prismmodelchecker.org/papers/fmsd-ptas.pdf. Accessed 25 May 2023
4. Sutalungka, T., Chariyathitipong, P., Vatanawood, W.: Transforming probabilistic timed automata to PRISM model. In: 26th International Computer Science and Engineering Conference (ICSEC), pp. 281–286. IEEE, Sakon Nakhon (2022)

5. PRISM Manual. https://www.prismmodelchecker.org/manual/Main/Introduction. Accessed 25 May 2023
6. Kumar, J.A., Vasudevan, S.: Formal probabilistic timing verification in RTL. Trans. Comput.-Aided Design Integr. Circuits Syst. **32**(5), 788–801 (2013)
7. Liu, Y., He, C.: A heuristics-based incremental probabilistic model checking at runtime. In: 11th International Conference on Software Engineering and Service Science (ICSESS), pp. 355–358. IEEE, Beijing (2020)

Fuzzy AHP-Based Evaluation of Key Success Factors in Digital Marketing for the Food Retail Industry

Sirorat Wiwatkajornsak(✉) and Chayathach Phuaksaman

Industrial Engineering, King Mongkut's University of Technology North, Bangkok 10800, Thailand
iaommy.sirorat@gmail.com

Abstract. This research explores an innovative methodology employing Fuzzy Analytical Hierarchy Process (FAHP) to evaluate the key success factors influencing digital marketing effectiveness within the food retail industry. In a dynamic market environment where digital strategies are integral to organizational success, understanding and prioritizing critical success factors is imperative. Traditional evaluation methods often struggle to capture the inherent uncertainties associated with multifaceted dimensions such as social media engagement, content relevance, and campaign performance. Leveraging FAHP, this study presents a comprehensive and adaptive framework designed to address the complexities of the digital marketing landscape. The proposed methodology incorporates fuzzy logic and linguistic variables into the AHP model, accommodating the inherent vagueness and subjectivity in expert judgments. Through the lens of FAHP, to identify, rank, and quantify the significance of key success factors in digital marketing for food retail. The approach not only provides a structured decision-making framework for industry practitioners but also enhances our understanding of the dynamic relationships between these factors. By bridging the gap between theory and practice, this research contributes valuable insights to the academic discourse and equips food retail businesses with tools to optimize their digital marketing strategies, decision-making processes, and foster competitiveness in an evolving market environment.

Keywords: Digital marketing · Food retail · Fuzzy AHP · Fuzzy logic · Decision making · Success factors in digital marketing

1 Introduction

The Food Retail Business is a crucial component of retail activities, playing a significant role in the economic system and impacting the economy of Thailand significantly. Its contribution to the overall service sector is substantial, accounting for approximately 4.7% and consistently expanding each year, playing a vital role in fostering the economic growth of Thailand. This business encompasses a diverse range of formats, from small-scale SMEs to large-scale retail outlets, including modern and varied street food

P. Meesad et al. (Eds.): IC2IT 2024, LNNS 973, pp. 73–84, 2024.
https://doi.org/10.1007/978-3-031-58561-6_8

establishments offering a wide array of products. Furthermore, consumer awareness and technological advancements have a substantial impact on the food retail business, necessitating adaptation to these changing circumstances. Information technology plays a pivotal role in enhancing various processes across the supply chain, utilizing information management systems for data storage, analysis, and decision-making with high-quality data. Particularly in the digital realm, elements such as digital marketing, the development of online businesses (e-commerce), and the use of technology to improve operational efficiency (business services) relevant to business operations have become crucial. In response to changing consumer purchasing behaviors, the industry is adjusting with a focus on convenience and the utilization of technology. This shift aims to create satisfying and high-quality consumer purchasing experiences by enabling rapid access to information and products. As a result, the food retail business can enhance its competitiveness efficiently.

The contemporary food retail landscape has undergone a transformative shift with the rise of digital marketing. In an era marked by rapid technological advancements and evolving consumer behaviors, digital marketing has become an indispensable tool for businesses navigating the intersection of tradition and the digital revolution. This transformation is particularly significant for the food retail industry, where digital marketing plays a pivotal role in shaping consumer perceptions, influencing purchasing decisions, and fostering brand loyalty.

Digital marketing, with its diverse online channels and strategies, has emerged as a dynamic and interactive platform for food retailers to engage with their target audience in real-time. From social media campaigns to personalized email marketing, these tools offer unprecedented opportunities to connect with consumers on a deeper level, bridging the virtual and physical shopping experience.

Beyond mere promotion, digital marketing in the food retail sector serves as a multifaceted instrument that amplifies brand visibility and adapts to evolving consumer preferences. As consumers increasingly rely on digital platforms for product discovery, price comparison, and reviews, food retailers must strategically leverage digital marketing to remain competitive. Recognizing this imperative, this research critically examines the key success factors underpinning effective digital marketing strategies in the food retail industry.

The study employs the Fuzzy Analytic Hierarchy Process (FAHP) as a methodological lens to scrutinize and prioritize these factors, offering a systematic and comprehensive approach to unravel the intricacies of digital marketing success by interview with experts in the retail food business, comprising owners of various food establishments in Thailand. The participants include those engaged in fine dining, cafeteria, buffet, bistro, and online retail food businesses that have incorporated digital marketing strategies into their operations. All five participants collectively possess a minimum of 10 years of experience in the food industry. In an environment where traditional and digital spheres coalesce, the efficacy of digital marketing strategies becomes crucial for success. This research aims to provide actionable insights to empower businesses in refining their strategies, enhancing customer experiences, and fortifying their position in a digitally driven marketplace. Through a meticulous exploration of the interplay between technology, consumer behavior, and marketing strategies, the study contributes valuable knowledge to the discourse surrounding the future of digital marketing in the food retail industry.

2 Literature Review

In the dynamic landscape of the global food retail industry, the intersection of technology and marketing strategies has become a cornerstone for sustainable growth and competitive advantage. As consumer behaviors evolve, driven by technological advancements and digitalization, food retailers are compelled to adopt innovative approaches to navigate the complex realm of digital marketing.

Background: The Dynamic Food Retail Industry
The food retail industry has long been recognized as a dynamic and consumer-driven sector, characterized by the constant evolution of consumer preferences and the ever-changing landscape of market trends. This dynamism demands agility from food retailers who must adapt to shifting consumer behaviors, regulatory changes, and the emergence of new competitors. The imperative for food retailers today is not merely to provide sustenance but to orchestrate immersive and personalized experiences that resonate with the expectations of a diverse and discerning customer base.

Digital Marketing in the Food Retail Sector
The advent of the digital era has catalyzed a paradigm shift in the way food retailers conceptualize and execute their marketing strategies. Digitalization does fundamentally impact firms' strategy development. With the fusion of IT and business strategy, Digital Business Strategy (DBS) creates the foundation for digital business models [1]. Digital marketing has transcended traditional boundaries, extending beyond physical storefronts to online platforms where consumers increasingly navigate their purchasing journeys. The evolution of digital marketing in the food retail sector is marked by a departure from conventional advertising to a more immersive, interactive, and data-driven approach that aligns with the preferences of today's tech-savvy consumers.

Food Retail Digital Marketing Operations
Digital marketing operations in the food retail sector are no longer confined to basic online presence or occasional promotions. Instead, they have evolved into sophisticated ecosystems encompassing various facets of online engagement. Social media channels, e-commerce platforms, mobile applications, and influencer collaborations all contribute to a comprehensive digital marketing strategy that spans the entire customer lifecycle. From building brand awareness to nurturing customer loyalty, digital marketing operations have become integral to the overall business strategy of food retailers. Through the strategic use of digital media, SMEs could build brand awareness and to leverage a global image, since the findings helped estimate a potential from branding perspectives. It is apparent that an SME will be able to increase brand awareness internationally if, through customization digital marketing strategies, it creates added value to its customers [2].

Technology and Data-Driven Strategies
The seamless integration of technology and data-driven strategies lies at the heart of

successful digital marketing operations in the food retail industry. Artificial intelligence, machine learning, and big data analytics have emerged as indispensable tools for extracting meaningful insights from the vast troves of data generated through online interactions. Digital marketing communication is important for businesses to offers opportunities for targeted engagement, measurable results, and building meaningful relationships with customers, contributing to overall business success and growth. Effective marketing communication will not only impact the progress of individual businesses but also foster the growth of Micro, Small, and Medium Enterprises MSMEs with enhanced competitiveness [3]. The timing of posts on social media is crucial for maximizing their impact and engagement. Optimal posting times is deemed crucial for devising an effective social media marketing strategy [4]. This technological integration empowers food retailers to understand consumer behaviors, preferences, and trends with unprecedented granularity, enabling them to tailor their marketing efforts for maximum impact.

Sales and Customer Experience

At the core of digital marketing in the food retail sector are two intertwined imperatives: driving sales and enhancing customer experiences. Customer satisfaction is the key for retail business successes. The significant impact of social media reviews and service operations on customer satisfaction [5]. Successful digital marketing strategies not only entice customers to make purchases but also prioritize delivering seamless and personalized experiences. Customer engagement is a strategic imperative for businesses to sustained growth, profitability, and brand perception. Digital engagement strategies serve as purposeful planning elements for implementing social media marketing capability, providing an overall approach to guide actions [6]. From user-friendly online platforms to targeted promotions based on individual preferences, the goal is to create a synergy between sales growth and customer satisfaction, fostering long-term relationships with the brand. Online platforms and e-commerce applications has witnessed significant enhancements in shipping, cost, and time, contributing to the increased effectiveness of the supply chain in the food retail industry [7]. Digital payment has become increasingly important in the modern financial landscape. Consumers also find it easy to make transactions with the option of payment by transfer because there is no need for meetings with buyers during the current New Normal Era [8].

Marketing Technology Usage in Food Retail

The deployment of marketing technology, including Search Engine Optimization (SEO), where explicit variable costs are challenging to attribute [9], marketing automation platforms, and analytics tools, has become a distinguishing feature of innovative food retailers. The positive effects stemming from the market leader's heightened SEO activities suggest potential positive externalities or spillover effects of digital marketing between firms [10]. These technologies streamline marketing operations, automate repetitive tasks, and provide actionable insights, enabling food retailers to stay ahead of market trends and consumer expectations. A strategic marketing approach focused on creating and distributing valuable, relevant, and consistent content to attract and retain a clearly defined audience and, ultimately, to drive profitable customer action [11]. The strategic usage of marketing technology is not merely a technological investment but a pivotal driver of competitive advantage in an industry where agility and responsiveness are paramount.

Business Performance Enhancement Through Digital Marketing
While digital marketing in the food retail industry has witnessed significant advancements, the ultimate measure of success extends beyond conventional metrics such as website traffic or social media engagement. Business performance enhancement is the goal, encompassing a spectrum of outcomes, from increased sales and revenue growth to heightened customer satisfaction and loyalty. Social media marketing variables on consumers' purchasing decisions, providing insights into the effectiveness of these marketing strategies [12]. The product and brand specific social media campaigns have more influence on business outcomes [13]. The holistic evaluation of digital marketing success factors is, therefore, a multifaceted undertaking that demands a nuanced and adaptive approach.

In recognition of this complexity, this article focuses on the integration of FAHP to evaluate the key success factors in food retail digital marketing. The subsequent sections delve into the methodology, application, and implications of employing FAHP in this context, offering a comprehensive understanding of how this approach can contribute to enhanced decision-making in the ever-evolving landscape of the food retail industry.

3 Methodology

Fuzzy AHP Method
The Fuzzy Analytic Hierarchy Process (FAHP) is a decision-making methodology that extends the classical Analytic Hierarchy Process (AHP) to handle imprecise information inherent in real-world scenarios. At its core lies the concept of fuzzy sets, a mathematical framework introduced by Lotfi Zadeh. Fuzzy sets allow the representation of uncertainty by assigning degrees of membership to elements, expressing shades of gray rather than rigid true/false distinctions. In the context of FAHP, fuzzy sets are employed to model the subjective judgments of decision-makers regarding the pairwise comparisons of criteria and alternatives.

The structure of FAHP involves several key steps. Firstly, decision criteria are identified, and pairwise comparisons are made to establish their relative importance. Fuzzy linguistic variables, such as "high", "medium", and "low", capture the imprecision in these comparisons. Subsequently, a fuzzy consistency check is applied to ensure the reliability of the obtained fuzzy comparison matrices. The final step involves aggregating these matrices to derive a comprehensive fuzzy priority vector, aiding decision-makers in reaching informed conclusions amidst uncertainty. The academic aim of employing FAHP is to provide a robust and flexible decision-making framework that accommodates the inherent vagueness and imprecision present in complex real-world problems. Researchers utilize FAHP to contribute to the advancement of decision sciences and provide valuable insights for practical applications across various domains.

The implementation of FAHP involves several delineated steps.

- **Step 1,** a hierarchical assessment questionnaire is formulated, and decision-makers' judgments are translated into fuzzy triangle numbers.

- **Step 2,** In instances where expressing judgments with fuzzy triangles proves challenging, decision-makers have the option to convey preferences using linguistic variables, from which equivalent fuzzy triangular values can be readily derived. Subsequently (see Table 1 and Fig. 1).
- **Step 3,** after establishing the hierarchy and conducting pairwise comparisons of alternative criteria, the computation of the global priority weight for each alternative is undertaken.

Table 1. Fuzzy comparison measures.

Linguistic terms	Triangular fuzzy numbers
Equally Important	1, 1, 1
Equally to Moderately Important	1, 2, 3
Moderately Important	2, 3, 4
Moderately to Strong Important	3, 4, 5
Strongly Important	4, 5, 6
Strongly to Very Strongly Important	5, 6, 7
Very Strong Important	6, 7, 8
Very Strongly to Extremely Important	7, 8, 9
Extremely Important	8, 9, 9

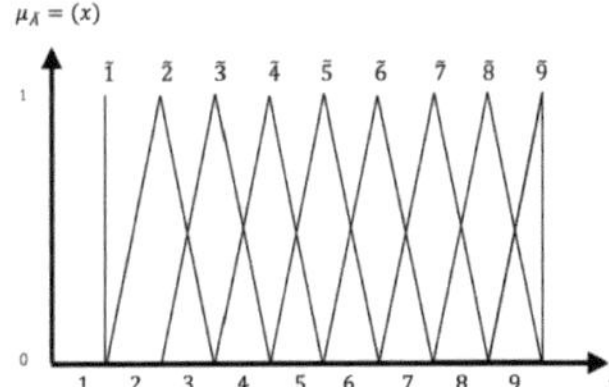

Fig. 1. Scale of fuzzy numbers.

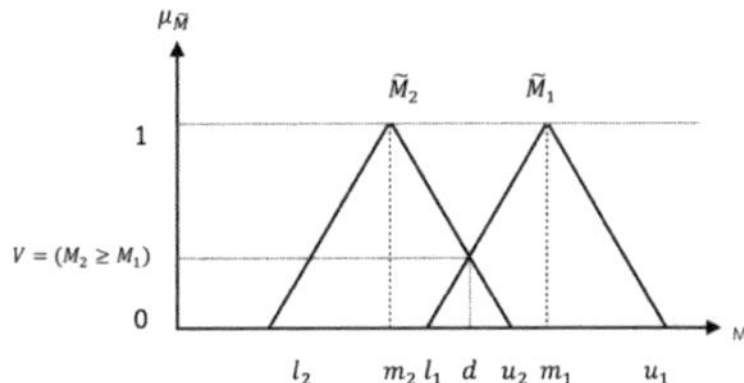

Fig. 2. Degree of possibility.

Problem Statement

The success of the food retail industry in the digital era hinges upon a comprehensive evaluation of four critical criteria. Firstly, Marketing and Brand Strategy play a pivotal role, encompassing three key factors: operation, foresight and vision, and business performance enhancement [14]. Operational efficiency, forward-thinking vision, and strategic enhancements to business performance collectively contribute to the formulation and execution of an effective marketing and brand strategy. Secondly, the Digital Marketing Strategy is a multifaceted domain, embracing four integral factors: sales and customer experience, culture and leadership, digital partnership, and digital marketing technique. The intertwining dynamics of these factors drive the overarching success of digital marketing initiatives within the food retail sector. Thirdly, Digital Technology

Usage stands as a crucial success factor, facilitating seamless operations and enhancing customer experiences through innovative technological applications. Lastly, Analytics and Data Analysis serve as the cornerstone, providing invaluable insights into consumer behavior, market trends, and campaign effectiveness. Together, these four criteria Marketing and Brand Strategy, Digital Marketing Strategy, Digital Technology Usage, and Analytics and Data Analysis constitute a holistic framework for gauging and optimizing the success factors in the digital marketing landscape of the food retail industry. This comprehensive approach acknowledges the nuanced interplay of strategic, technological, and analytical components, underscoring the multifaceted nature of success in the digital realm.

The implementation of key success factors in digital marketing for the food retail industry involves several delineated steps.

Pairwise Comparison and Calculation of Weights and Ranking

During first step, the comparison on first level between key success factors criteria. On the second level, the pairwise comparison between sub-criteria.

Matrix Comparison

$$\tilde{A} = \left(\tilde{a}_{ij}\right)_{n\times n} = \begin{bmatrix} (1,1,1) & (l_{12}, m_{12}, u_{12}) & \cdots & (l_{1n}, m_{1n}, u_{1n}) \\ (l_{21}, m_{21}, u_{21}) & (1,1,1) & \cdots & (l_{2n}, m_{2n}, u_{2n}) \\ \vdots & \vdots & \vdots & \vdots \\ (l_{n1}, m_{n1}, u_{n1}) & (l_{n2}, m_{n2}, u_{n2}) & \cdots & (1,1,1) \end{bmatrix} \quad (1)$$

then

$$\tilde{a}_{ij} = \left(l_{ij}, m_{ij}, u_{ij}\right) = \tilde{a}_{ij}^{-1} = \left(\frac{1}{u_{ji}}, \frac{1}{m_{ji}}, \frac{1}{l_{ji}}\right) for\ i, j = 1, \ldots, n\ and\ i \neq j \quad (2)$$

For the normalized summation of the rows in the matrix, the following Eq. (3)

$$S_i = \sum_{j=1}^{m} M_{gi}^{j} \otimes \left[\sum_{i=1}^{n}\sum_{j=1}^{m} M_{gi}^{j}\right]^{-1} \quad (3)$$

M_{gi}^{i} is the triangular fuzzy numbers in pairwise matrices and can be calculated from Eq. (4), (5), (6)

$$\sum_{j=1}^{m} M_{gi}^{j} = \left[\sum_{j=1}^{m} l_j \sum_{j=1}^{m} m_j \sum_{j=1}^{m} u_j\right]^{-1} \quad (4)$$

and

$$\left[\sum_{i=1}^{n}\sum_{j=1}^{m} M_{gi}^{j}\right]^{-1} = \left(\frac{1}{\sum_{i=1}^{n} u_i}, \frac{1}{\sum_{i=1}^{n} m_i}, \frac{1}{\sum_{i=1}^{n} l_i}\right) \quad (5)$$

and

$$\sum_{i=1}^{n}\sum_{j=1}^{m} M_{gi}^{j} = \left(\sum_{i=1}^{m} l_i \sum_{i=1}^{m} m_i \sum_{i=1}^{m} u_i\right) \tag{6}$$

Degree of Possibility of $S_i \geq S_j$, , (see Fig. 2).
$S_i = (l_i, m_i, u_i), S_j = \left(l_j, m_j, u_j\right); i \neq j$ from Eq. (7)

$$V\left(S_i \geq S_j\right) = \begin{cases} 1 & m_i \geq m_j \\ 0 & l_j \geq u_i \\ \frac{(l_j - u_i)}{(m_i - u_i) - (m_j - l_j)} & other \end{cases} \tag{7}$$

Convex Fuzzy Number.
$S_i \geq S_j$ then $i \neq j$ as shown in Eq. (8)

$$S_i \geq S_j | j = 1, 2, \ldots, m; i \neq j = minV(M \geq M_i | i = 1, 2, 3 \ldots, k) \tag{8}$$

Weight Vector as shown in Eq. (9)

$$W_i' = minV\left(S_i \geq S_j | j = 1, 2, \ldots, m; i \neq j\right) \tag{9}$$

W' as shown in Eq. (10)

$$W' = \left(W_1', W_2', \ldots, W_n'\right)^T \tag{10}$$

The defuzzified weight need to be normalized by using Eq. (11)

$$W_i = \frac{W_i'}{\sum_{i=1}^{n} W_i'} \tag{11}$$

$$W = (W_1, W_2, \ldots, W_n)^T \tag{12}$$

Calculating the Consistency Index (CI) and the Consistency Ratio (CR) of the comparison matrix where λmax is the largest eigenvalue of the comparison matrix. CR should be ≤ 0.1 the judgement matrix is satisfied and RI is the random index by using the following Eq. (13), (14)

$$CI = \frac{(\lambda max - n)}{n - 1} \tag{13}$$

The consistency ratio (Table 2)

$$CR = \frac{CI}{RI} \tag{14}$$

Table 2. Ratio index (RI) for different value of n.

n	1	2	3	4	5	6	7	8	9	10
RI.	0	0	0.52	0.89	1.12	1.25	1.34	1.4	1.45	1.49

Table 3. The summary of the main criteria of the key success factors.

Main Criteria	Normalised Weight	Ranking	CR
Marketing and Brand Strategy	0.565	1	0.036
Digital Marketing Strategy	0.256	2	
Digital Technology Usage	0.116	3	
Analytics and Data Analysis	0.062	4	

Table 4. The summary of the sub-criteria of the key success factors.

Main Criteria	Sub-criteria	Normalised Weight	Ranking	CR
Marketing and Brand Strategy	Operation	0.718	1	0
	Business Performance Enhancement	0.185	2	
	Foresight and vision	0.097	3	
Digital Marketing Strategy	Sales & Customer Experience	0.414	1	0.07
	Digital Marketing Technique	0.228	2	
	Digital Partnership	0.228	3	
	Culture and leadership	0.130	4	
Digital Technology Usage	Marketing Technology Usage	0.633	1	0.082
	Social Media Monitoring	0.144	2	
	Website Traffic	0.110	3	
	Web User Experience	0.018	4	
	Digital Payment	0.032	5	
	Marketing Channel	0.026	6	

(continued)

Table 4. (*continued*)

Main Criteria	Sub-criteria	Normalised Weight	Ranking	CR
	Mobile Technology	0.015	7	
	Machine Learning	0.011	8	
	Web analysis	0.009	9	
	Result Measurability	0.002	10	
Analytics and Data Analysis	Marketing Research	0.361	1	0.03
	Difference between platform	0.115	2	
	Social Media Capability	0.103	3	
	Purchase decision	0.082	4	
	Type of activity	0.077	5	
	Time of day	0.075	6	
	Social media communication	0.067	7	
	Day of the week	0.061	8	
	Environment Condition	0.058	9	

Table 5. The summary of the sub-criteria 2 of the key success factors.

Sub-criteria	Sub-criteria 2	Normalised Weight	Ranking	CR
Business Performance Enhancement Pandey et al. (2020)	ROI	0.296	1	0.073
	Interactivity	0.227	2	
	Value Co-Creation	0.189	3	
	Improved buyer-Supplier	0.162	4	
	Sustainable	0.085	5	
	Marketing Performance	0.043	6	
Sub-criteria	Sub-criteria 2	Normalised Weight	Ranking	CR
Sales and Customer Experience	Customer target	0.308	1	0.025
	Brand Awareness	0.228	2	
	Customer Engagement	0.187	3	

(*continued*)

Table 5. (*continued*)

Sub-criteria	Sub-criteria 2	Normalised Weight	Ranking	CR
	Customer Retention	0.171	4	
	Customer acquisition	0.105	5	
Digital Partnership	Co-Partner	0.341	1	0.035
	Influencer Marketing	0.338	2	
	Affiliate program	0.171	3	
	Sponsors	0.150	4	
Digital Marketing Technique	Content Marketing	0.375	1	0.052
	Social Media Advertising	0.213	2	
	E-mail Marketing	0.122	3	
	SEO	0.109	4	
	SEM	0.109	5	
	Digital Marketing Budget	0.071	6	
Environment Condition	Innovation technology change	0.856	1	0
	Competition in market	0.144	2	

4 Results and Discussion

The result of the key success factors in digital marketing for the food retail industry (see Table 3 and Table 4) (Table 5).

Conclusions

Employing Fuzzy AHP to assess key success factors in digital marketing for the food retail industry reveals a nuanced hierarchy. Marketing and Brand Strategy, Digital Marketing Strategy, Digital Technology Usage, and Analytics and Data Analysis emerge as pivotal elements. The prioritization, informed by Fuzzy AHP, underscores the intricate interdependencies among operational efficiency, business performance enhancement, sales and customer experience and technological adeptness insights. This integrated approach enables the food retail sector to navigate the complexities of the digital landscape, fostering a dynamic equilibrium that optimally enhances customer experiences, ensures sustainable growth, and fortifies market competitiveness.

References

1. Holotiuk, F., Beimborn, D.: Critical success factors of digital business strategy. In: 13th International Conference on Wirtschaftsinformatik, Switzerland, pp. 991–1005 (2017)
2. Makrides, A., Vrontis, D., Christofi, M.: The gold rush of digital marketing: assessing prospects of building brand awareness overseas. Bus. Perspect. Res. **8**(1), 4–20 (2019)
3. Rusdana, N.R., Choirani, S.J., Friska, A.: Digital marketing communication strategy for micro, small and medium enterprises (MSMEs) in business competition. Int. J. Res. Appl. Technol. **2**(1), 163–168 (2022)
4. Devereux, E., Grimmer, L., Grimmer, M.: Consumer engagement on social media: evidence from small retailers. J. Consum. Behav. **19**(2), 151–159 (2019)
5. Ramanathan, U., Subramanian, N., Parrott, G.: Role of social media in retail network operations and marketing to enhance customer satisfaction. Int. J. Oper. Prod. Manag. **37**(1), 105–123 (2017)
6. Drummond, C., O'Toole, T., McGrath, H.: Digital engagement strategies and tactics in social media marketing. Eur. J. Mark. **54**(6), 1247–1280 (2020)
7. Foster, B.: New era chain supply on retail food industry in Indonesia. In: IOP Conference Series, vol. 1115, no. 1, p. 012019 (2021)
8. Darma, G.S., Noviana, I.P.T.: Exploring digital marketing strategies during the new normal era in enhancing the use of digital payment. J. Mantik **4**(3), 2257–2262 (2020)
9. Erdmann, A., Casado, J.M.P.: Digital inbound marketing: measuring the economic performance of grocery e-commerce in Europe and the USA. Technol. Forecast. Soc. Chang. **162**, 120373 (2021)
10. Casado, J.M.P., Erdmann, A.: E-commerce customer attraction: digital marketing techniques. Evolution and dynamics across firms. J. Promot. Manage. **27**(5), 697–715 (2021)
11. Olson, E.M., Olson, K.M., Czaplewski, A.J., Key, T.M.: Business strategy and the management of digital marketing. Bus. Horiz. **64**(2), 285–293 (2021)
12. EtAl, H.E.: Effect of digital marketing and social media on purchase intention of SMES food products. Turk. J. Comput. Math. Educ. **12**(3), 3672–3678 (2021)
13. Dolega, L., Rowe, F., Branagan, E.: Going digital? The impact of social media marketing on retail website traffic, orders and sales. J. Retail. Consum. Serv. **60**, 102501 (2021)
14. Pandey, N., Nayal, P., Rathore, A.S.: Digital marketing for B2B organizations: structured literature review and future research directions. J. Bus. Industr. Mark. **35**(7), 1191–1204 (2020)

A Rule-Based Expert System for Automated Document Editing

Sandeep Varma, Shivam Shivam, Soumya Deep Roy(✉), and Biswarup Ray

ZS Associates Pune, Pune, India
{sandeep.varma,shivam.shivam,soumya.b.roy,biswarup.ray}@zs.com

Abstract. This research paper introduces a Rule-Based Expert System designed for the automated editing of documents in PDF and PPT formats. The system employs a set of predefined rules, extracted from guideline documents using a Large Language Model (LLM), to execute tasks such as redaction of sensitive text/logo detection and annotation of text elements that deviate from prescribed font size guidelines. Following the detection and annotation process, the system further enhances documents by resizing the detected text elements based on the predefined rules. To achieve these editing tasks, the system integrates advanced image processing techniques, leveraging fine-tuned Optical Character Recognition (OCR) for accurate text extraction from document images. Furthermore, Natural Language Processing (NLP) algorithms are utilized to analyze and interpret textual content. The combination of image processing, OCR, NLP, and rule extraction using LLM ensures a comprehensive approach to document editing, enhancing efficiency and accuracy. The proposed system addresses the need for automated and rule-driven document editing, contributing to advancements in information security and document standardization.

Keywords: Rule-based Expert System · Image processing · Optical Character Recognition · Document Sanitization · Natural Language Processing

1 Introduction

In an era characterized by the incessant exchange of vast volumes of data among diverse entities, the imperative to safeguard sensitive information remains paramount. The exchange of classified data, spanning client information in corporations, patient medical records in the healthcare sector, to confidential government documents, necessitates meticulous data protection measures. With the escalating volume of shared documents, the conventional approach to data protection through manual sanitization becomes a laborious and time-consuming endeavour, posing the risk of inadvertent omissions.

This paper introduces a pioneering solution to this challenge - a Rule-Based Expert System for Automated Document Editing. Our system is designed to

P. Meesad et al. (Eds.): IC2IT 2024, LNNS 973, pp. 85–94, 2024.
https://doi.org/10.1007/978-3-031-58561-6_9

address the growing need for efficient and accurate document sanitization, particularly in the context of PDF and PPT formats. By leveraging pre-trained open-source LLM, advanced image processing techniques, fine-tuned OCR, and NLP algorithms, our system automates the redaction of sensitive information and the identification of non-compliant text elements. In addition to redaction and detection, the proposed model enhances the document by performing the rescaling of detected text regions to the desired size using image processing. This groundbreaking approach not only enhances the speed and accuracy of the sanitization process but also mitigates the risk associated with manual efforts, contributing significantly to the evolution of secure and streamlined document management practices.

This research paper is structured as follows: Sect. 2 reviews recent research on automatic document editing. Section 3 provides an overview of the dataset and details the proposed model. Experimental results are presented and discussed in Sect. 4, and the paper concludes in Sect. 5 with insights into prospects and potential enhancements.

2 Literature Survey

In this section, we discuss foundational research works on optical character recognition and image inpainting, which are integral components of our proposed pipeline. Additionally, we address research works related to document sanitization. Image inpainting is employed to remove distortions, such as texts, blocks, lines, noise, or various forms of masks, and recent algorithms can reinstate both the structure and texture components of an image. Zeng et al. [1] proposed the utilization of the Saliency Map and Gray entropy for image inpainting. For object removal from images, Zhang et al. [2] utilized a joint probability density function to implement image inpainting. Optical Character Recognition currently is widely used for text recognition, to extract the text from documents, images, etc. OCR is majorly divided in 6 steps - image acquisition, preprocessing, segmentation, feature extraction, classification, and post-processing [3]. A further 7-step approach was also proposed [4] by Chaithanya et al. Sankaran et al. [5] discussed a Bidirectional LSTM-based framework for OCR to operate at the word level. Zhang et al. [6] proposed an iterative Conditional Random Field (CRF) to separate content and non-content parts and apply OCR on the content part. In their paper [7], Varma et al. used pytesseract ocr for extracting multimodal information and built a video indexing system on top of it. Tian et al. [8] fused CNNs with RNNs and enhanced the accuracy of horizontal text with the help of a vertical anchor method in their Connectionist Text Proposal Network.

3 Proposed Methodology

The proposed methodology encompasses a structured pipeline for document editing and enhancement. Beginning with the ingestion pipeline, documents undergo extraction of slides/pages, followed by image pre-processing. Key to the methodology is the extraction of rules from a guideline document utilizing LLM. Subsequently, the system employs Keras OCR and inpainting algorithm for the redaction of text and logos based on the extracted rules. The process includes the detection of textual elements that violate guidelines, with a subsequent step for the rescaling of detected text regions to the desired size. The high-level design of the proposed pipeline is shown in Fig. 1.

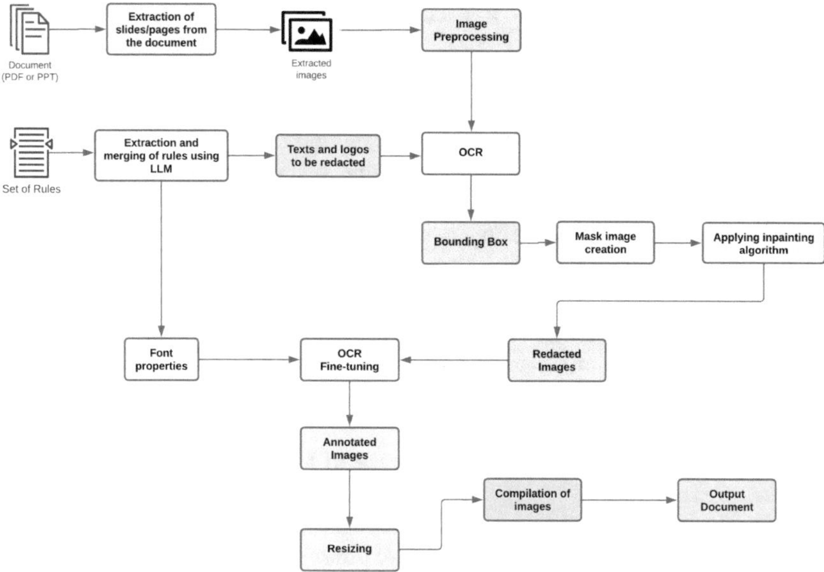

Fig. 1. Proposed pipeline.

3.1 Dataset Description

The dataset for this study is meticulously curated by aggregating documents of diverse formats, including PowerPoint and PDF. Specifically, we focused on hand-selecting documents rich in graphical content to comprehensively evaluate the efficacy of our proposed model. Our in-house dataset comprises a total of 60 documents, 20 PDF files, and 20 PowerPoint files. These documents are sourced from a variety of publicly available organizational materials on the internet, and obtained from reputable platforms such as Scribd, Slideshare, and various other credible sources. The deliberate inclusion of documents from different formats and reputable sources aims to enhance the robustness and real-world applicability of our evaluation, ensuring a thorough examination of our proposed model's effectiveness across diverse document types.

3.2 Document Ingestion

The input document is processed through the ingestion module, where pages or slides are extracted as images based on the file extension. The python-pptx library is employed to develop a robust extraction pipeline for PowerPoint (PPT) documents. With its functionalities, we programmatically access and manipulate PPT files, enabling the extraction of slides and their conversion into images. For PDF documents, PyMuPDF has been utilized to open the PDF document and access individual pages, we efficiently render each page as an image for further analysis. The library's versatility enhances our ability to navigate through documents and integrate this functionality into our broader processing framework, contributing to the overall efficiency and adaptability of our system for handling PDF documents. Subsequently, these images are stored in a designated folder, each tagged with its respective page/slide number. After extraction, the images undergo comprehensive pre-processing using a set of image processing techniques, including re-scaling, binarization, de-skewing, noise removal, and thresholding. These pre-processing methods are implemented to enhance OCR performance. The seamless integration of the extraction pipeline with the image pre-processing pipeline has significantly enhanced the efficiency and versatility of our document ingestion system.

3.3 Extraction of Rules

The guideline document undergoes parsing, and the extracted texts are fed into the open-source Llama-2-7b-chat-hf large language model (LLM) for summarization and rule extraction. An open-source Large Language Model (LLM) is employed to safeguard the confidentiality of guideline documents. The LLM processes the text, leveraging advanced natural language processing capabilities, to derive concise summaries. Simultaneously, it identifies rules related to font properties, sensitive text elements, and images earmarked for redaction. Utilizing the summarization output, the LLM extracts key information, facilitating a more focused analysis in subsequent stages. The extracted rules (as shown in Fig. 2) serve as guidelines for the redaction process, ensuring a systematic and accurate approach to handling diverse document elements. This integration of parsing, summarization, and rule extraction contributes to a comprehensive and effective document processing pipeline.

3.4 Redaction of Texts and Logos

The extracted images are fed into a pre-trained Keras OCR module, which detects and extracts all the text elements along with their corresponding bounding box coordinates. The extracted terms from the LLM are then cross-referenced with the text obtained through Keras OCR. Matched terms and their corresponding bounding box coordinates are systematically stored in an array. This consolidated array serves as a map for precise redaction, facilitating the seamless identification and removal of specific textual content within the images.

index	Parameters	value
0	permissible font-size	20
1	permissible font-weight	5
2	permissible pixel-height	50
3	permissible pixel-length	112
4	restricted text(s)	["This information is classfied","Confidential ingredients","atc-tech","searching index"]

Fig. 2. LLM extracted rule parameters

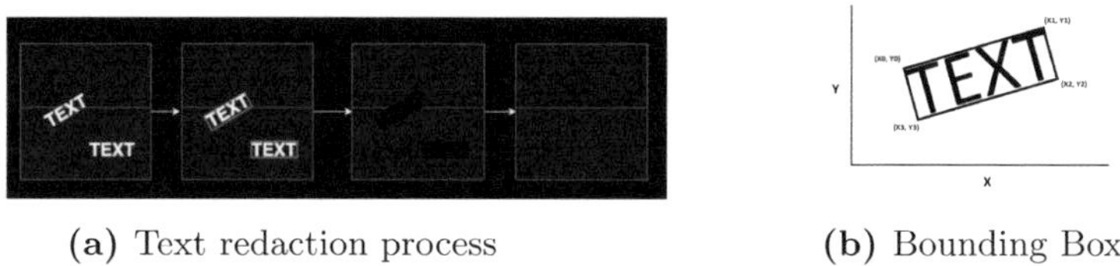

(a) Text redaction process (b) Bounding Box

Fig. 3. Text redaction process-flow

Each of the annotated images, along with the matched terms and their corresponding bounding box coordinates, is then passed into the next module for mask image creation. A mask is an image composed of binary pixels, with non-zero and zero values. The mask must share the same dimensions as the input image. Non-zero pixels in the mask correspond to regions of interest containing the input text that will undergo inpainting, while regions with zero pixels won't be reconstructed after applying the inpainting algorithm (as depicted in Fig. 3a). In this study, line masks are utilized for their flexibility in covering detected texts with varying orientations. To apply the mask, it is essential to calculate the thickness of the line and determine the coordinates of the starting and ending points of the line using the bounding box axes. This involves evaluating the midpoint between the bottom-left and top-left corners for the starting point and the midpoint between the top-right and bottom-right corners for the ending point. The calculated thickness of the line is crucial for the proper application of the mask in the redaction process. The thickness of the line is determined by measuring the length of the line that connects the top-left corner axis and the bottom-left corner axis. The mathematical formula for this measurement is provided below:

$$thickness = \sqrt{(x2 - x1)^2 + (y2 - y1)^2} \tag{1}$$

The starting point of the line is determined by calculating the midpoint between the bottom-left corner axis (x3, y3) and the top-left corner axis (x0, y0) of the bounding box (as shown in Fig. 3b). Similarly, the endpoint of the mask line is established as the midpoint between the top-right corner axis (x1, y1) and the bottom-right corner axis (x2, y2). Equation 2 is employed to calculate the midpoint between two coordinates. The coordinates of the midpoints are calculated using the following formula:-

$$x_{mid} = \frac{(x_i + j_j)}{2} \tag{2}$$

After the creation of the mask image, the original image is inpainted utilizing Opencv's cv2.INPAINT_NS function. In case an image contains matched texts multiple times, the image will undergo inpainting as many times as the frequency of the detected text. Figures 4b and 5b show how multiple masks are used to inpaint an image. The annotated images are then stored serially in a designated folder.

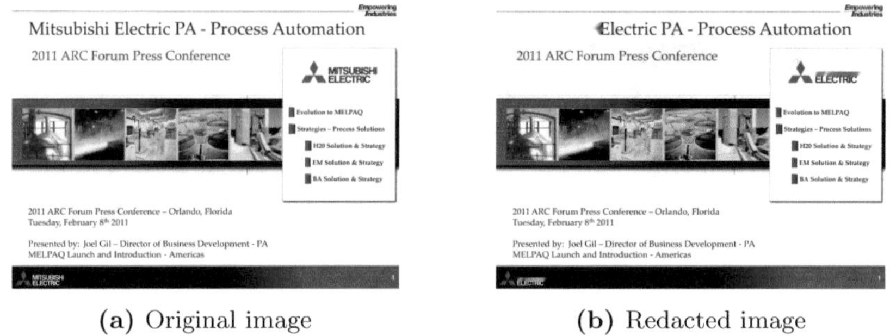

(a) Original image **(b)** Redacted image

Fig. 4. Inpainting an image using multiple masks (Here, the term 'Mistubishi' is redacted from the image)

(a) Original image **(b)** Redacted image

Fig. 5. Inpainting an image using multiple masks (Here, the terms 'Airtel' and 'Amazon Prime Video' are redacted from the image)

3.5 Detection of Textual Elements Violating the Guidelines

For subsequent detection and annotation methodology, the redacted images are fed into the fine-tuned Keras model, equipped with the extracted rules. This model selectively extracts text regions of a specific height, creating bounding boxes around the text regions that violate the prescribed threshold (as shown in Fig. 6b and 7b).

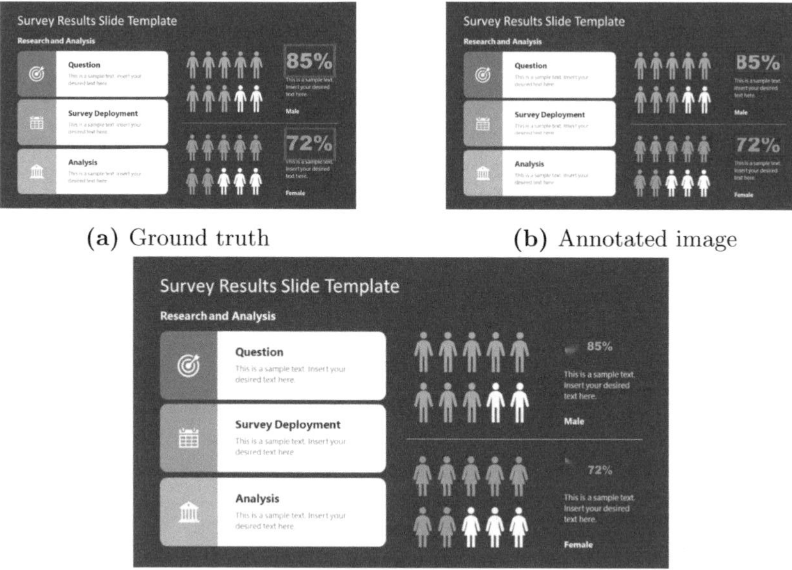

(a) Ground truth

(b) Annotated image

(c) Output image

Fig. 6. In (a), the red bounding boxes mark the text regions violating the guidelines. In (b), the blue bounding boxes are annotated by our proposed model

3.6 Rescaling of Detected Text Regions to the Desired Size

For the subsequent task of rescaling the identified text to a desired size, the process unfolded with the extraction of the image enclosed by the boundary box. This image was then resized in accordance with a specified height. To seamlessly integrate the resized text back into the original context, a masking step was introduced using the OpenCV Inpaint Function. This function facilitated the adjustment of the masking color to seamlessly blend with the background. The final step involved overlaying the resized and masked image onto the original image, with precise positioning achieved through calculated adjustments to the coordinates (as shown in Fig. 6c and 7c). Specifically, the new coordinates (new_x1, new_y1) were computed by considering the differences in height (h_diff) and width (w_diff) between the original and resized text. This comprehensive methodology ensures the effective fine-tuning and annotation of text regions in the research paper's context.

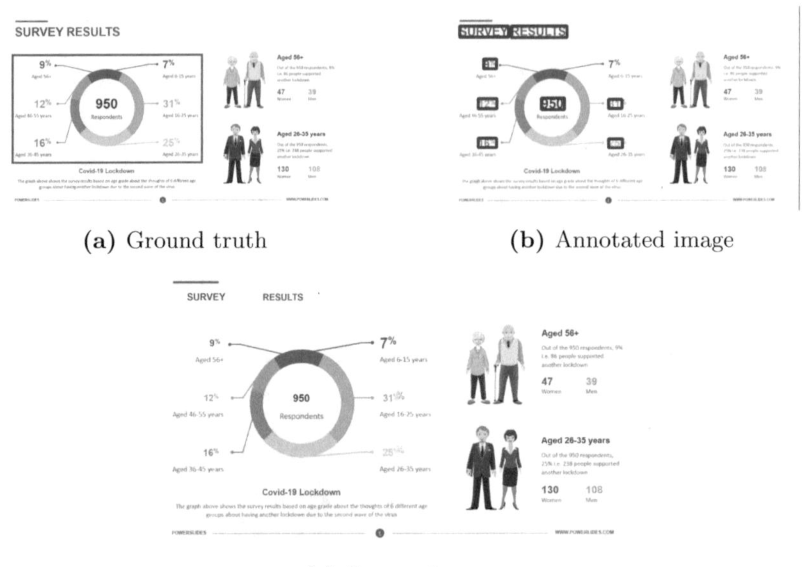

(a) Ground truth

(b) Annotated image

(c) Output image

Fig. 7. In (a), the red bounding boxes mark the text regions violating the guidelines. In (b), the blue bounding boxes are annotated by our proposed model

3.7 Compiling Images into the Output Document

In the final stage of the pipeline for compiling images into PDF and PowerPoint (PPT) files using Python, redacted images are compiled to form the final output document. This comprehensive process leverages libraries like PyPDF2 for PDFs and python-pptx for PowerPoint. Images undergo systematic processing, formatting, and integration into their respective file formats, ensuring an efficient and automated solution for streamlined document generation.

4 Result and Analysis

To assess our model, we employed the recall metric. Recall, or sensitivity, is a metric that evaluates the number of correct positive predictions made out of all potential positive predictions. Mathematically, the Recall score could be expressed by the following formula:-

$$Recall = \frac{TruePositive}{TruePositive + FalsePositive} \tag{3}$$

In the context of our study, for all images containing sensitive elements, recall informs us about the proportion of these images that the model successfully identified as sensitive information. We compared the original images (ground truth) with the reconstructed images to assess the degree of redaction performed by our model. We juxtaposed each extracted image with the corresponding reconstructed image to determine the effectiveness of our model in successfully redacting sensitive elements within an image (as enumerated in Table 1).

Recall was also utilized to evaluate how well the model performed in detecting and annotating textual elements that violate the prescribed font properties in a document (as shown in Table 2).

Table 1. Calculated recall scores across various document formats

	Highest Recall score	Lowest Recall Score	Average Recall Score
Powerpoint Documents	1.000	0.846	0.902
PDF Documents	0.909	0.766	0.823

Table 2. Computed recall metrics

	Highest Recall score	Lowest Recall Score	Average Recall Score
Powerpoint Documents	0.989	0.823	0.934
PDF Documents	0.952	0.833	0.900

After thorough research and analysis, we identified the factors responsible for the low recall scores:

1. Low pixel density of the image
2. Detected texts are skewed or curved to a large extent
3. Erroneous detection of texts using OCR
4. Mask creation is not covering the region of interest(s)
5. Inpainting applied to the image is not up to mark

5 Conclusion and Future Enhancements

In conclusion, the Rule-Based Expert System for Automated Document Editing presented in this research offers a promising avenue for addressing the escalating challenges associated with safeguarding sensitive information in shared documents. By automating the process of sanitization and editing, the system not only enhances efficiency but also mitigates the inherent risks of manual interventions. The integration of image processing, fine-tuned OCR, and NLP provides a comprehensive solution, ensuring the accurate redaction of sensitive information and the identification of non-compliant text elements.

Looking ahead, future enhancements to this system could leverage the advancements in Generative AI models. Fine-tuning these models to the specific requirements of document sanitization could usher in a new era of automated intelligence. Generative AI, with its ability to understand context and generate human-like text, holds immense potential for refining and automating the decision-making process in document editing. This would not only expedite the editing process but also improve the adaptability of the system to evolving document structures and content types.

References

1. Zeng, J., Fu, X., Leng, L., et al.: Image inpainting algorithm based on saliency map and gray entropy. Arab. J. Sci. Eng. **44**, 3549–3558 (2019). https://doi.org/10.1007/s13369-018-3592-5
2. Zhang, D., Liang, Z., Yang, G., et al.: A robust forgery detection algorithm for object removal by exemplar-based image inpainting. Multimedia Tools Appl. **77**, 11823–11842 (2018). https://doi.org/10.1007/s11042-017-4829-0
3. Mittal, R., Garg, A.: Text extraction using OCR: a systematic review. In: 2020 Second International Conference on Inventive Research in Computing Applications (ICIRCA), Coimbatore, India, pp. 357–362 (2020). https://doi.org/10.1109/ICIRCA48905.2020.9183326
4. Chaithanya, C.P., Manohar, N., Issac, A.B.: Automatic text detection and classification in natural images. Int. J. Recent Technol. Eng. (IJRTE) **7**, 176–180 (2019)
5. Sankaran, N., Jawahar, C.V.: Recognition of printed Devanagari text using BLSTM neural network. In: Proceedings of the 21st International Conference on Pattern Recognition (ICPR 2012), Tsukuba, Japan, pp. 322–325 (2012)
6. Zhang, H., Liu, C., Yang, C., Ding, X., Wang, K.: An improved scene text extraction method using conditional random field and optical character recognition. In: 2011 International Conference on Document Analysis and Recognition, Beijing, China, pp. 708–712 (2011). https://doi.org/10.1109/ICDAR.2011.148
7. Varma, S., Pandey, A., Shivam, Das, S., Roy, S.D.: Video indexing system based on multimodal information extraction using combination of ASR and OCR. In: Sachdeva, S., Watanobe, Y., Bhalla, S. (eds.) BDA 2021. LNCS, vol. 13167, pp. 201–208. Springer, Cham (2022). https://doi.org/10.1007/978-3-030-96600-3_14
8. Tian, Z., Huang, W., He, T., He, P., Qiao, Y.: Detecting text in natural image with connectionist text proposal network. In: Leibe, B., Matas, J., Sebe, N., Welling, M. (eds.) ECCV 2016. LNCS, vol. 9912, pp. 56–72. Springer, Cham (2016). https://doi.org/10.1007/978-3-319-46484-8_4

Physical and Mental Support System for Older Adults Using Communication Robot

Kazuo Hemmi[1(✉)], Aiko Mieno[2], and Tae Yamaguchi[2]

[1] Department of Information Systems, University of Nagasaki, Nagasaki, Japan
hemmi@sun.ac.jp

[2] Department of Nursing Science, University of Nagasaki, Nagasaki, Japan
{aikoie,tae0813}@sun.ac.jp

Abstract. People aged 65 and over are generally referred to as older adult, and these people often require physical and mental support. Therefore, we developed a physical and mental support system for the older adults by using the communication robot, Unibo. In daily physical support, it is important to be able to obtain vital values quickly and easy to check them, and in mental support, communication is important. Hence the system measures body temperature, blood pressure, and pulse rate and provides physical support for contacting medical institutions about the measurements and mental support for reducing loneliness and anxiety through conversations and games. The experimental results demonstrated that the system has the potential to relieve loneliness and anxiety in the older adults while aiding medical institutions in obtaining measurement data directly. Moreover, the system was demonstrated to be a friendly verbal communication interface for older adults.

Keywords: Older Adults · One-person Household · Health Care · Human Interface · Communication Robot

1 Introduction

The aging society is increasing in population globally. The World Health Organization defines an older adult as a person above the age of 65 [1]. According to the United Nations, the proportion of the older adults will increase from 2020 to 2050 in all parts of the world, and the world's population aged 65 and above is expected to increase from 9.3% to 16.0% [2]. The ratio of the older adults in Japan is relatively high; according to the Cabinet Office, the aging rate was 28.4% (35.8 million people) in 2019, and it is expected to reach 37.7% in 2050 [3].

To solve the aging problem, various healthcare devices and older adults life support robots have been developed for home medical care/nursing care and general support. These devices utilize information and communication technology. Particularly, several studies have been conducted to monitor the health conditions of the older adults living alone by using a monitoring system based on physical health data to prevent lonely deaths [4–7]. Although advances in information equipment make it easy to build a system that provides physical support, little has been done in the advancement of information

P. Meesad et al. (Eds.): IC2IT 2024, LNNS 973, pp. 95–104, 2024.
https://doi.org/10.1007/978-3-031-58561-6_10

equipment for reducing loneliness, anxiety and improving the mental health of the older adults. However, these few studies suggest that communication robots can be potential partners to improve the mental of the older adults. In a previous study about the mental effect of communication robots on the older adults live in the nursing home, experiments were conducted for five weeks by using seal robots [8].

Libin et al. reported that humans achieve positive emotions, such as joy and interest, through their involvement with interactive robots that have the appearance of living things [9]. Exploratory pilot tests using a conversation agent-based system also showed that agents actively engaging older people in interactions are more effective against loneliness than those of passive conversations [10]. These results indicate that a communication robot can reduce the feeling of loneliness among the older adults. While several studies have investigated the relationship between the quality of life of the older adults and the use of support robots, research has not been conducted on a system that simultaneously provides physical and mental care for the older adults. Therefore, we aimed to develop a system that supports the physical and mental care of older adults aged 65 and above. Thus, we developed a support system that uses a communication robot named Unibo, and we evaluated the system from the subjective and objective perspectives of the participants and on the speech recognition rate, respectively. In this system, the communication robot regularly measures the body temperature, blood pressure, and pulse of the older adults. The results are then sent to medical personnel to provide physical care support and provide conversations and games to alleviate loneliness and anxiety.

2 Methods

2.1 System Description

The communication robot is the core of the proposed system, and Unibo (Unirobot Corp.) was used because of its functionality and appearance familiarity. As shown in Fig. 1, Unibo has a deformed childlike appearance. The robot is 32 cm in height, 26 cm in width, 16 cm in depth, and 2.5 kg in weight. Unibo picks up the speech sound of the participant through the microphone on the head, and the speech is analyzed by the application programming interface (API) of the conversation on the cloud via the Internet. The API returns an appropriate response according to the semantic content of the analyzed speech. The robot's face is a liquid crystal display (LCD) touch screen that reads the touch operation of the participant and displays facial expressions or images. A camera is attached above the face, and it can take pictures and record videos. Both arms of the robot move 90° back and forth by a shaft motor, and touch sensors are mounted on the legs. The robot has a built-in communication module that enables wireless communication such as Wi-Fi and Bluetooth.

Our research on an oral function improvement system for the older adults [11] has shown that, in many cases, the tablet terminal does not respond to the touch operations of the older adults. Therefore, in this study, the communication mode of the system is mainly verbal, and the touch operation is a hybrid specification that is performed when necessary. The physical support system that measures vital signs does not require any touch operation, and the mental health support system mainly uses verbal conversations and instructions, unless occasionally during quizzes that require voice or tap operations

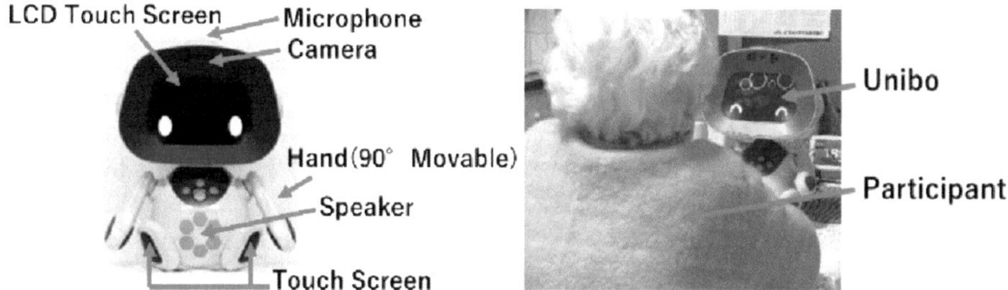

Fig. 1. The communication robot, Unibo.

Fig. 2. A participant using Unibo.

to select answers. In this study, the distance between Unibo and the participant was set at approximately 1 m. Figure 2 shows an example of a participant using Unibo.

In this study, the physical support system for measuring vital signs aims to support the management of the daily physical conditions of the older adults. Unibo gives verbal instructions to older adults so that it can measure vital values using a thermometer and sphygmomanometer. Unibo automatically collects the acquired results and sends them to the medical personnel. This system is designed to enable older adults to naturally measure body temperature and blood pressure by listening to the verbal instruction from Unibo and watching the images displayed on Unibo's screen. The activity diagram of this system is shown in Fig. 3. Unibo automatically collects the measurement results from each device and completes the daily report by sending an e-mail summarizing the measurement data to the medical personnel.

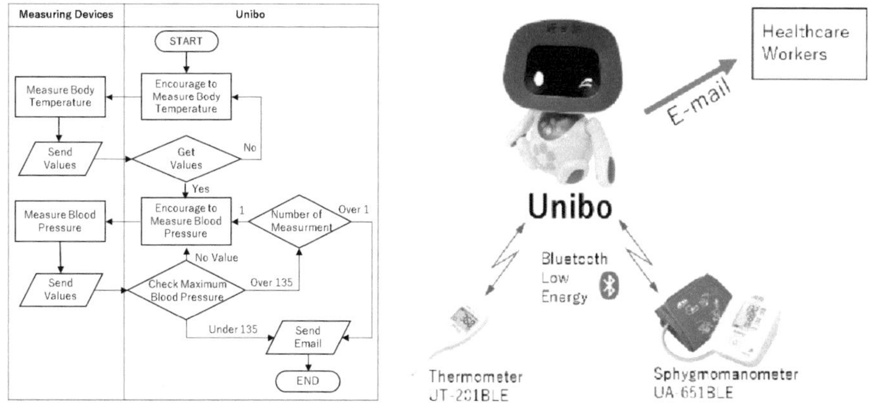

Fig. 3. Flow chart of the vital signs measurement system.

Fig. 4. System configuration diagram.

The vital sign information measured by the physical support system are blood pressure, pulse rate, and body temperature because abnormality in these vital signs can be easily detected. The devices used for these measurements were selected based on compatibility with Unibo. Thus, an oscillometric-type electronic sphygmomanometer connected to the upper arm (UA-651BLE, A&D Co., Ltd,) was used, and a thermistor-type electronic thermometer connected to the axilla (estimated 30 s) (UT-201BLE, A&D

Co., Ltd.) was used. The results measured by each device are sent to Unibo by wireless communication, and Bluetooth low energy is used as the wireless communication standard. Figure 4 illustrates the configuration of the vital sign support system for evaluating the physical conditions.

To increase the speech recognition rate of Unibo, depending on the situation, we used two speech recognition functions. The first function is a mode that starts speech recognition when the older adults says "Unibo" (hereafter "wake-up word"). The flow chart of this mode is shown in Fig. 5. In this mode, Unibo only responds when it detects a wake-up word. Therefore, even if Unibo were installed in the living space of the older adults, it would not interfere with unnecessary reactions. After speaking the wake-up word to Unibo, it is expected to become easier for the older adults to identify the time when Unibo's speech recognition begins. Therefore, the trigger mode is expected to increase the success rate of speech recognition when calling various contents activated by speaking to Unibo, such as the beginning of a conversation with Unibo games, and Internet searching. Hence, the trigger mode was made the default setting when installing Unibo in the experimental environment as well as in the experiment of calling various contents and evaluating the impression of the contents. Unibo can engage in continuous and flexible daily conversation ("free conversation") with the participant. However, in these situations, it is complicated and cumbersome to speak the word of awakening each time the older adults speaks. Therefore, to realize continuous and flexible free conversations, the trigger mode setting is disabled.

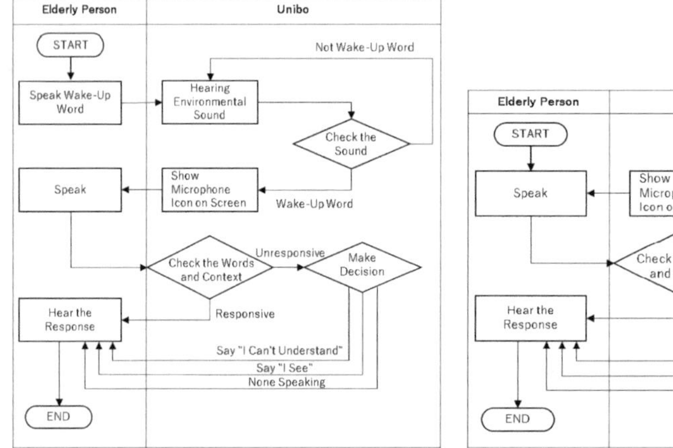

Fig. 5. Flow chart of the trigger mode.

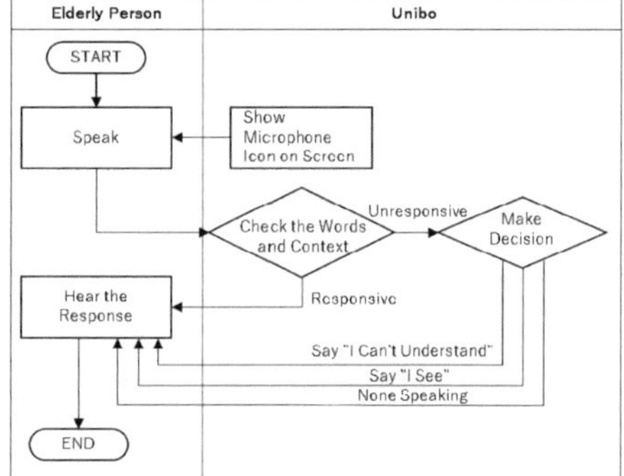

Fig. 6. Flow chart of the non-trigger mode.

Figure 6 is an activity diagram for when the trigger mode is turned off. During a free conversation, for example, if a older adult utters the statement, "Good morning", to off? Unibo, the robot will respond likewise by saying "Good morning. Did you sleep well yesterday?" The older adults will then respond appropriately to the question. Unibo again responds with appropriate reactions and expressions to the speeches of the elderly. The conversation continues this way, establishing free conversations. This flexibility of free conversations is a major conversational feature of the developed system.

2.2 Survey Method

The participants in this study are older adults.We aimed to evaluate the system through subjective evaluation by the participants and objective evaluation by checking the speech recognition rate. Hence, the survey was divided into three sections to verify the applicability of the system. Additionally, the experimental scenery were record by a video camera.

In the first section, participants evaluated the physical support system subjectively by testing the system that measures vital signs at the verbal instruction of Unibo. Here, the participant operated the thermometer and sphygmomanometer while following Unibo's instructions for measuring vital signs. The experiment was completed when the measurement results were sent by e-mail to the designated terminal.

In the second section, the participant activated Unibo's contents using the trigger mode and experienced the activated content. This helped to obtain the impression for the subjective evaluation; the objective evaluation for checking the effectiveness of the trigger mode was performed by observing the videos of the section. Here, the investigator first verbally explained the use of the trigger mode. The participant then activated two types of contents using the trigger mode: the song and quiz contents. This section ended when the execution of the two types of contents was completed.

In the third section, the participant and Unibo engaged in a free conversation with the trigger mode turned off. The participant obtained the impression for subjective evaluation in the free conversation with Unibo, and the recorded videos of this section were used for the objective evaluation to determine the effectiveness of the free conversation. During the experiment, the investigator urged the participant to greet Unibo or interview Unibo first and afterward ask Unibo freely. This section ended after three minutes.

After completing the three sections, the participant subjectively evaluated the system by filling a questionnaire containing prepared questions and free-form fields. The questionnaire items were classified into two parts: a part collected the attributes of the participant, and the other part collected the subjective evaluation responses of the system. The participants were instructed to complete the VAS (Visual Analogue Scale) questionnaire before and after the experiment to assess changes in mental state. The scale possessed scores of 0–10 ("0 = dark feeling" to "10 = bright feeling"), and then marking the current feeling on the line (see Fig. 7).

Fig. 7. Example of VAS.

The experimental scenes were recorded by turning on the video camera from behind the participants. The videos were recorded to protect privacy. In addition, the effectiveness of each speech recognition rate in the trigger and non-trigger modes was confirmed by checking the recorded videos. This was done by counting the number of times in which communication was not established between Unibo and the participant in each mode and dividing the count by the total number of conversations in each mode ("error rate").

This study was conducted with the approval of the University of Nagasaki General Research Ethics Committee with approval number 440.

3 Result

3.1 Participant's Characteristics

The attributes of the participants are shown in Table 1. The participants included 7 male and 4 females. Their ages ranged from 65 to 88 years, with a mean ± SD of 71.1 ± 6.7. No participant had a disability in the upper limbs. All participants had experience with mobile devices, including tablets, but none had experience with communication robots.

Table 1. Characteristics of participants.

Age, y	
Mean ± SD	71.1 ± 6.7
Range	75–90
Sex	
Male	7 (63.6%)
Female	4 (36.4%)
Are there any motor dysfunctions (especially dominant arms) in the upper limbs (and fingers)?	
No	11 (100%)
Have dysfunctions but is able to operate the system	0 (0%)
Disable to operate the system	0 (0%)
Experience using mobile terminals (smartphones and tablets)	
Yes/No	11 (100%)/0 (0%)
Experience using communication robot	
Yes/No	0 (0%)/11 (100%)

3.2 Evaluation of the System

The results of evaluating the system operation are shown in Table 2. The impression about the system was extremely good for 4 participants (36.4%) and moderately good for 7 (63.6%). The ease of operation during the temperature measurements was extremely easy for 7 participant (63.6%), moderately easy for 2 (18.2%), moderately difficult for 2 (18.2%). The ease of operation during the blood pressure measurements was extremely easy for 9 The ease of operation during the temperature measurements was extremely easy for 7 participant (63.6%), moderately easy for 2 (18.2%), moderately difficult for 2 (18.2%). (81.8%), moderately easy for 2 (18.2%). The ease of operation during the body temperature measurements was extremely easy for 7 participant (63.6%), moderately easy for 2 (18.2%), moderately difficult for 2 (18.2%). All 11 (100%) faced no problem with the touch function of Unibo's screen. The conversation with Unibo was extremely fun for 5 participants (45.5%), moderately fun for 6 (54.5%). In response to the question "Do you think this system can reduce loneliness and anxiety?", 5 participants (45.5%) said able and 6 (54.5%) said moderately able. In response to the question "Do you think this system can reduce loneliness and anxiety by contacting medical institutions?", 5 participants (45.5%) said able and 6 (54.5%) said moderately able. When asked if they would like to use the proposed system in the future, 10 participants (90.9%) said yes, whereas 1 (9.1%) said no. The results of mental state before and after the experiments are shown in Table 3. The average and ± SD of the VAS values before the experiment was 6.1 cm ± 1.9 cm, and after the experiment it was 8.3 cm ± 1.0 cm; the rate of change average ± SD was 28.1% ± 17.0%. Table 4 presents the result of each mode and the experimental error rate. The average and ± SD of the error rate in the trigger mode non-trigger mode and total of both mode was 26.5% ± 26.8, 28.0% ± 14.9, and 27.4% ± 21.2, respectively.

4 Discussion

In this research, we developed a system that supports the physical and mental of the older adults using a conversation robot and verified the system. It can be said that the developed system is friendly with the older adults because all the participants evaluated the system impression as more than moderately good.

The average ± SD of the mental state changes before and after the experiment measured by VAS was 28.1% ± 17.0. This indicates that the participants became cheerful after experiencing the developed system. Hence, it can be said that the proposed system, which uses Unibo's verbal communication interface, is highly suitable for the older adults. Additionally, the system can alleviate the anxiety and loneliness of the older adults effectively by sending the measured values of vital signs to the medical institution. Another remarkable result was that all participants reported no problem when operating Unibo's LCD by touch, and the LCD was easy for them to operate.

In the measurement of vital signs using each measurement device, 2 participants said as moderately hard, meaning there is room for improvement. This was because the timing of the thermometer's Bluetooth communication did not match that of Unibo, making some of the participants repeat the measurements. In addition, it was sometimes difficult for the participants to wrap the sphygmomanometer's manchette on their upper arms by themselves. Hence, the measuring device and Unibo should be adjusted to ensure that the measurement is performed in one attempt, and the sphygmomanometer should be changed to a model that users can easily use themselves.

By using the trigger mode during the experiment, a clear improvement in speech recognition was observed for participants who understood the trigger mode specifications. However, no improvement in the speech recognition rate was observed for the participants who did not understand the specifications of the trigger mode and the timing of the start of speech recognition by Unibo. This observation is also reflected in the standard deviation of the error rate in the trigger mode, with an average error rate ± SD of 26.5% ± 26.8. Further, it was found that the average error rate in the trigger mode

Table 2. The results of evaluating the system operation.

How do you feel about this system?	
Extremely Good Impression	4 (36.4%)
Moderately Good Impression	7 (63.6%)
Moderately Bad Impression	0 (0%)
Moderately Bad Impression	0 (0%)
Was the operation easy during body temperature	
Extremely Easy	7 (63.6%)
Moderately Easy	2 (18.2%)
Moderately Hard	2 (18.2%)
Extremely Hard	0 (0%)
Was the operation easy when measuring blood pressure?	
Extremely Easy	9 (81.8%)
Moderately Easy	2 (18.2%)
Moderately Hard	0 (0%)
Extremely Hard	0 (0%)
Did the touch operation problem?	
Yes/No	0 (0%)/11 (100%)
Did you enjoy the conversation with Unibo?	
Extremely Fun	5 (45.5%)
Moderately Fun	6 (54.5%)
Moderately Not Fun	0 (0%)
Extremely Not Fun	0 (0%)
Do you think this system can reduce loneliness and anxiety?	
Able	5 (45.5%)
Moderately Able	6 (54.5%)
Moderately Unable	0 (0%)
Unable	0 (0%)
Do you think this system can reduce loneliness and anxiety by contacting medical institutions?	
Able	5 (45.5%)
Moderately Able	6 (54.5%)
Moderately Unable	0 (0%)
Unable	0 (0%)
Do you want this system for use in your home?	
Yes/No	10 (90.9%)/1 (9.1%)

Table 3. The results of mental state by VAS.

	Average Score	Standard Deviation
Before Experiment	6.1 cm	1.9
After Experiment	8.3 cm	1.0
Rate of Change	28.1%	17.0

Table 4. Confirmation of error rate.

Characteristics of Participants		Average Score	Standard Deviation
n = 11, Age = 71.1 ± 6.7 Range of Age = 65–88	Trigger Mode	26.5%	26.8
	Non-Trigger Mode	28.0%	14.9
	Total of Both Mode	27.4%	21.2

was slightly reduced compared with that in the free conversation. As the error rate in the free conversation was 28.0% ± 14.9, the SD suggests that the difference between the participants was not larger than in trigger mode. Hence, it can be said that the trigger mode improved the voice recognition rate depending on the understanding of the specifications. The considerable dependence of the voice recognition on the participant's understanding of the specifications suggests that the trigger mode needs to be improved to make the specifications easier to understand.

Several comments were written in the comment section. These comments consist of positive comments and negative comments (see Table 5). The positive comments were about the robot's cuteness and high level of perfection. Our research group have conducted research for the older adults using tablets [11]. Through this research, we learned the lesson that "devices must be cute" If the device is not cute, it will be difficult to get it accepted by the older adults. Negative comments were related to Unibo's conversational ability. As noted above, the participants' comments also indicate that the Unibo's conversational skills need to be improved.

Table 5. Comments from participants.

Positive comments	Negative comments
• Unibo is cute • Voice is cute • This robot attains a high level of perfection • It was fun because of unexpected questions and answers	• Conversation response is slow • Robot doesn't respond well • Conversation timing with Unibo is difficult

5 Conclusions and Future Works

In this study, we developed and verified the usability of a system that supports the physical conditions and mental health of older adults aged 65 and above by using a communication robot named Unibo. The verification results demonstrated that the system has a human interface that the older adults consider to be friendly, and it was found that the older adults can speak with Unibo in the same way that young people use the system. In addition, the questionnaire responses suggested that the system has the same effect of reducing anxiety and loneliness as the support provided by medical institutions. Regarding the speech recognition rate, the results differed considerably depending on how well the participant understood the system specifications, making it necessary to improve the usability of the interface.

References

1. World Health Organization: Men ageing and health. https://apps.who.int/iris/bitstream/handle/10665/66941/WHO_NMH_NPH_01.2.pdf;jsessioni. Accessed 17 Dec 2023
2. United Nations: World population ageing 2020 highlights. https://www.un.org/development/desa/pd/sites/www.un.org.development.desa.pd/files/undesa_pd-2020_world_population_ageing_highlights.pdf. Accessed 17 Dec 2023
3. Cabinet Office of Japane: Annual report on the ageing society [summary] FY 2020, https://www8.cao.go.jp/kourei/english/annualreport/2020/pdf/2020.pdf. Accessed 3 Mar 2023, 21 Dec 2023
4. Takami, S., Torii, I., Ishii, N.: Development of system for prevention of solitary death with mobile devices. Procedia Comput. Sci. **35**, 1193–1201 (2014). https://doi.org/10.1016/j.procs.2014.08.216
5. Nguyen, T.H.: Brief paper: A study on the possibility of introducing Korean technologies into Vietnam for monitoring and prevention of solitary deaths of elderly. J. Multimed. Inf. Syst. **6**(1), 31–35 (2019). https://doi.org/10.33851/jmis.2019.6.1.31
6. Suzuki, M., Iduka, K.: A liveliness sensing system for solitary aged persons. In: 2018 7th International Congress on Advanced Applied Informatics (IIAI-AAI), Japan (2018). https://doi.org/10.1109/iiai-aai.2018.00209
7. Maki, H., Ogawa, H., Matsuoka, S., Yonezawa, Y., Caldwell, W.M.: A daily living activity remote monitoring system for solitary elderly people. Presented at the 2011 33rd Annual International Conference of the IEEE Engineering in Medicine and Biology Society (2011). https://doi.org/10.1109/iembs.2011.6091357
8. Wada, K., Shibata, T., Saito, T., Tanie, K.: Effects of robot-assisted activity for elderly people and nurses at a day service center. Proc. IEEE **92**(11), 1780–1788 (2004). https://doi.org/10.1109/jproc.2004.835378
9. Libin, A.V., Libin, E.V.: Person-robot interactions from the robopsychologists' point of view. The robotic psychology and robotherapy approach. Proc. IEEE **92**(11), 1789–1803 (2004). https://doi.org/10.1109/jproc.2004.835366
10. Ring, L., Barry, B., Totzke, K., Bickmore, T.: Addressing loneliness and isolation in older adults: Proactive affective agents provide better support. In: 2013 Humaine Association Conference on Affective Computing and Intelligent Interaction (ACII), pp. 61–66 (2013). https://doi.org/10.1109/acii.2013.17
11. Mieno, A., Hemmi, K., Nagamine, T., Yamasumi, N., Sakamoto, H., Yoshida, E.: Development of oral function improvement system using tablet computer. In: 2018 5th International Conference on Behavioral, Economic, and Socio-Cultural Computing (BESC), pp. 83–86 (2018). https://doi.org/10.1109/besc.2018.8697256

Prediction Cows Estrus Images Using Convolutional Neural Network with Optimized Parameters by the Artificial Immune System Algorithm

Watchara Ninphet, Nopadol Amm-Dee(✉), and Adisak Sangsongfa

Department of Industrial Technology Management, Faculty of Industrial Technology, Muban Chom Bueng Rajabhat University, Chom Bueng 70150, Thailand
{watcharanin,noppadolamd,adisaksan}@mcru.ac.th

Abstract. The factor that affects the milk production of dairy cows is the breeding of cows. Before the cows can be bred, the cows must show signs of estrus first. Farmers must detect estrus in time because the period of estrus is It did not take long, so this research has developed the prediction of estrus in dairy cows using artificial intelligence methods, that is, using convolutional neural networks to help in the prediction. This research has a Comparison between the convolutional neural networks before adjusting the parameters, which achieved an accuracy of 95.082% and when using the Artificial Immunity System algorithm to adjust the parameters, the accuracy was 98.361%. The prediction used 4 types of dairy cow movements for prediction.

Keywords: Convolutional Neural Network · Cow Estrus · Artificial Immunity System

1 Introduction

The dairy farming industry in Thailand has been continually promoted and supported by both the public and private sectors, but the amount of milk produced is still not sufficient to meet the demand for domestic consumption. Therefore, it is necessary to import powdered milk and other dairy products. An approach that can reduce production costs for dairy farmers is to increase the proportion of births to female dairy calves, which have a higher economic value than male calves. Because if female dairy calves can be produced with high potential, it will be possible to increase the amount of raw milk production, causing production costs for farmers to decrease as well [1] and [2].

However, as mentioned above, the importance of raising dairy cows to obtain milk for consumption. In producing dairy cows that will produce milk, one important factor is that the dairy cows must be pregnant having a calf will allow the cow to produce milk. In addition, the reproductive cycle of dairy cows begins with the onset of estrus. The reproductive cycle is the first time that dairy cows show estrus. It can be seen from 7–20-month-old dairy cows. Their readiness for breeding is taken into account. The fitness

P. Meesad et al. (Eds.): IC2IT 2024, LNNS 973, pp. 105–120, 2024.
https://doi.org/10.1007/978-3-031-58561-6_11

of the body, size, structure, and body weight is greater than the age of the dairy cow. The level of food received according to the season will be a factor affecting the onset of reproductive age [3]. Therefore, observing the symptoms of estrus in dairy cows, you can observe the following symptoms: 1) Restlessness and snuggling with other cows 2) Trying to ride other cows that are trying to escape (at this stage it is difficult to know which cow has affected estrus) 3) Cows during actual estrus will allow others to ride them calmly for a time suitable for mating. 4) The vulva is swollen, and the tail bobs slightly upwards. 5) There is a long stream of clear mucus flowing. Leaving the vagina stains the anal area, unlike in pregnant animals where the mucus is very sticky and 6) The vaginal mucosa is red because there is more blood supplied [4]. When dairy cows are in estrus, which usually occurs at night dairy farmers need to monitor estrus all the time because when dairy cows are in estrus, the appropriate time for artificial insemination is approximately 12 h after Breeder stands still and allows other cows to ride. Estrus is a process that requires a lot of time, knowledge, and attention from farmers. If you miss the opportunity, you will have to wait for another estrus period, which will result in losing the opportunity to mate. In management, there is research on this topic that has accomplished this monitoring with different methods, such as Serap Göncü, et al. [5], who presented a study titled, "The Sensor Technologies for More Efficient Cow Reproduction Systems", in which sensors are used to detect estrus in cattle. The researchers used three types of sensors in this study: electrochemical sensors and chemo resistive gas sensors. The researchers found that chemo resistive gas sensors were suitable and worked well, but the more useful ones were electrochemical sensors, but the problem was that they were large and impractical to use. However, the problem was that they were large and inconvenient to use. This was because the sensors used in the experiment had to be attached to the dairy cow's body, which is inconvenient and can easily damage the sensors. And many other research projects use various related instruments, such as those of Thi Thi Zin [6], Asmaa S. Ali, et al. [7], etc.

In addition to the research that uses various tools or technologies attached to the body of the dairy cow or the stall of the dairy cow as mentioned above, there is also research that uses other techniques to detect the estrus of the dairy cow. For example: Shogo Higaki et al. [8] presented a paper titled Estrus Detection Using Background Image Subtraction Technique in Tie-Stalled Cows, in which the researchers proposed a new method developed by combining the background subtraction technique and the thresholding method to detect spikes. of activity related to estrus in stationary cattle. Using this method A significant increase in estrous activity was detected. And the sensitivity for detecting estrus was as high as 90% with an accuracy of 50%, respectively. This result may indicate that activity tracking using background subtraction techniques has the potential to be a non-invasive method for detecting estrus. No contact in cattle is tied to gather very well. Zhen Wang et al. [9] uses machine learning to analyze images and present a study entitled A Non-Contact Cow Estrus Monitoring Method Based on the Thermal Infrared Images of Cows. The researcher proposes a noncontact method for monitoring estrus in cows. Based on thermal infrared images of cows A laboratory color space-based feature extraction method was proposed for the thermal infrared image of the cow's eye and vulva. LOGISTIC and SVM (Support Vector Machine) models were used to model the estrus of cattle using Thermal infrared temperature changes in estrus and non-estrus cattle. The experimental results show that the heat detection rate of the LOGISTIC-based model is 82.37% and the heat detection rate of the SVM-based model is 81.42% under the optimal segmentation profile. The maximum temperature in the cow's eyes and vulva is an input. And the recall rate is above 86%. In machine Learning, in the 21st century, most researchers prefer learning and testing learning with artificial neural networks because of the high accuracy of the output results. An example of this type of research would be Briefly summarized as follows: Juan Rodrigues Alvarez et al. [10] presented research that used convolutional neural networks to estimate the body condition score (BCS) of cows. The researchers significantly expanded the techniques that that system utilizes. Through the use of transfer learning and ensemble modeling techniques. To improve the accuracy of BCS estimation even further. The improved system achieved good estimation results compared to the base system. The overall accuracy of BCS estimates within 0.25 units of difference from the actual value increased by 4% (up to 82%), while the overall accuracy within 0.50 units increased by 3% (up to 97%). Alexey Ruchay et al. [11] use convolutional neural networks to predict suitable slaughter body shapes because, on large farms, precise and expensive industrial scales are used to measure live weight. Researchers have presented a new model for live weight prediction based on 3D cloud augmentation in the form of azimuth projection and deep learning image regression has been shown that in real data sets, the accuracy of weight measurement using the model which is presented as high as 91.6%.

Therefore, from the problems and importance as well as related research mentioned above, the researcher will present the use of convolutional neural networks with parameters optimized appropriately with the input values to predict the characteristics of estrus in dairy cows to obtain the most accurate output values by using CCTV equipment to take both still and moving images to monitor estrus. This will benefit dairy farmers and reduce the cost of investing in expensive equipment to detect estrus in dairy cows.

2 Literature Review

2.1 Convolutional Neural Network (CNN)

Before explaining convolutional neural networks, it is important first to understand the term artificial neural network. Artificial neural networks are a part of machine learning. And it is at the heart of deep learning algorithms. It consists of a hierarchy of nodes which consists of the input layer. One or more hidden layers and output layer Each node is connected to another node and has an associated weight and threshold. If the output of each node is above the specified threshold value, that node will be activated. It sends data to the next layer of the network. Otherwise, no data is sent to the next layer of the network. Although we mainly focus on feedforward networks, there are many different types of neural networks. They are used for different use cases and data types. For example, recurrent neural networks are commonly used for natural language processing and speech recognition. While convolutional neural networks (ConvNets or CNNs) are commonly used for classification and computer vision tasks, however, before CNN, functional feature extraction methods were used in the past. It takes a long time to identify objects in images.

How do convolutional neural networks work? Convolutional Neural Network, it is different from other neural networks. Due to their superior efficiency in ingesting data as audio, images, speech, or sound, they have three main types of layers:

- Convolutional layer
- Pooling layer
- Fully-connected (FC) layer

The convolutional layer is the first layer of a convolutional network. While additional convolutional layers or pooling layers can follow convolutional layers, the fully connected layer is the final layer. With each layer, the CNN increases in complexity, identifying greater portions of the image. Earlier layers focus on simple features, such as colors and edges. As the image data progresses through the layers of the CNN, it starts to recognize larger elements or shapes of the object until it finally identifies the intended object. Convolutional layer It is the first layer of the network. Convolutional Although the Convolutional layer can be followed by floor Convolutional additional or grouped floors the layer that is completely connected is the last. With each layer, the CNN increases in complexity. Previous layers focused on simple features like color and edges by specifying more parts of the image. As the image data progresses through the layers of the CNN, it starts to recognize larger elements or shapes of objects. Until finally being able to identify the desired object. Suppose that you want to calculate in CNN with Input size $3 \times 3 \times 1$ and Weight size $2 \times 2 \times 1$, the calculation steps are: Find the sum of the multiplication between Input and weight using the same set of weights and scan across the Input from left to right and top to bottom (1)–(4). The result (Z) is

$$Z_1 = W_1X_1 + W_2X_2 + W_3X_4 + W_4X_5 \tag{1}$$

$$Z_2 = W_1X_2 + W_2X_3 + W_35 + W_4X_6 \tag{2}$$

$$Z_3 = W_1X_4 + W_2X_5 + W_3X_7 + W_4X_8 \tag{3}$$

$$Z_4 = W_1X_5 + W_2X_6 + W_3X_8 + W_4X_9 \tag{4}$$

It will be noted that what is different from normal Neural networks is that the weight values used to calculate Z are all the same set of values. This is one reason why we use ConvNet because if we use a normal Neural network, the model will have to learn a lot of weight values. However, ConvNet consists of two layers: a Convolutional layer and a Pooling layer. Convolutional layer: Hyperparameters are weight window size, input scaling (Padding), and scanning steps (Stride). Pooling layer: Data scaling. The idea is that instead of using all the image data, it is divided into areas and then selects certain values from that area (similar to the way humans look, sometimes a glance at the image can tell you what it is.). In general, two types of Pooling layers can be used: 1) Max pooling: Select the highest value from the Pool size. 2) Average pooling: The average of the Pool size. Parameters that must be defined. (Hyper parameters) are scanning steps (Stride) and Pool size [12] (Fig. 1).

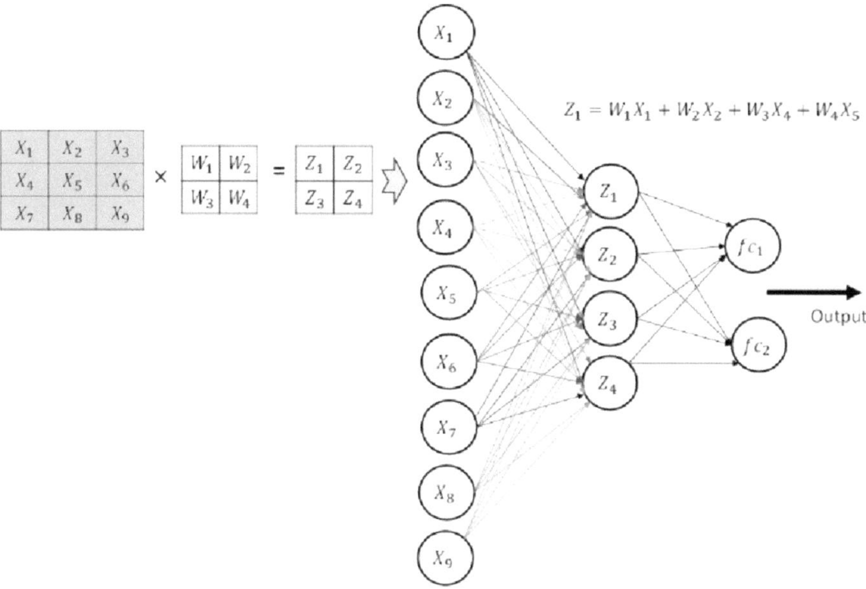

Fig. 1. The figure shows CNN working steps to write neural network model.

2.2 Artificial Immune Systems Algorithm (AIS)

Artificial immune systems are algorithms and systems that use the human immune system as inspiration. The human immune system is a robust, decentralized, error-tolerant, and adaptive system. Such properties are highly desirable for the development of new machine-learning systems. Unlike other bio-inspired techniques such as genetic algorithms and artificial neural networks, the field of AIS covers the spectrum of existing algorithms due to the different algorithms. Each uses different properties of different cells. All AIS algorithms mimic the behavior and properties of immune cells. Specifically B cells, T cells, and dendritic cells (DC), but the resulting algorithms show different levels of complexity and can perform a variety of tasks. At present, AIS has developed more than before. However, the well-known pseudocode of AIS can be shown in Fig. 2 as follows.

```
Initialize the value of AIS parameters {antibody population (P), iterations (I_max ), and percentage of
antibody elimination (%β)}.
Generate a population of P antibodies based on the feasible fleet of vehicles
For each antibody (i ∈ P) calculate affinity (i)
Set start iteration (I)=1
Do {
        For each antibody (i)
                Calculate the number of clones (N_c) and clone antibody (i)
                For each clone, apply inverse mutation to create a new antibody
                    Calculate the affinity of the new antibody from the tour obtained
                    If affinity (new antibody) is better than the clone then clone = new antibody
                    Else perform shift operation mutation to create a new antibody
                        Calculate the affinity of the new antibody from the tour obtained
                        If affinity (new antibody) is better than the clone then
                            clone= new antibody
                    antibody (i) =clone
        Eliminate the worst antibodies from the population on %β
        Create new antibodies to replace the eliminated antibodies
        I+= 1
}
While I ≤ I_max
```

Fig. 2. The figure shows the pseudocode of artificial immune systems algorithm.

3 Methodology

The research framework shown in Fig. 3 shows the sequence of work steps in each step-in detail. Each step can be described in order as follows.

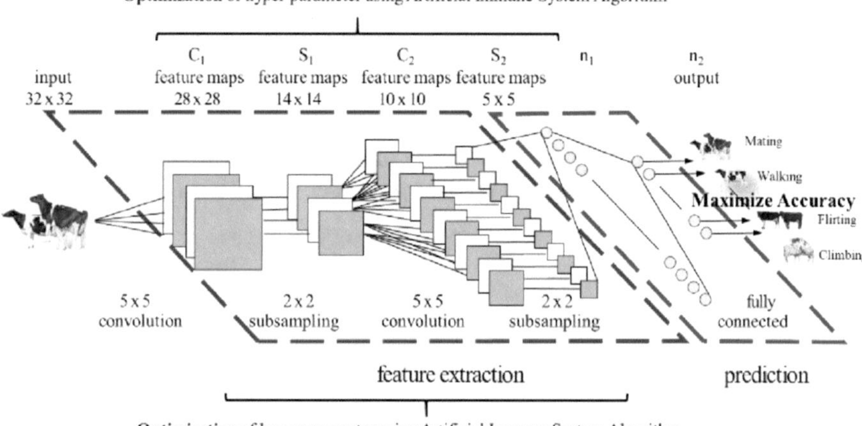

Fig. 3. The figure shows framework of prediction cows estrus images using convolutional neural network with optimized parameters by the artificial immune system algorithm.

3.1 Data Preparation

Collecting images from the process of collecting real-time video images and then making them into still images is divided into sub-steps as follows.

1. Steps for preparing cattle pens Installing the video camera Use dairy cows with a Holstein Friesian bloodline level of 87.50% or more, using 2nd grade or higher from a standard dairy farm certified by the Department of Livestock Development, 10 cows. The experiment was in a cow pen measuring 4 × 8 square. Meters, the roof is at least 4 m above the ground, and install an IP camera.
2. Real-time video image collection process, video image collection at a resolution of 1280 × 960 pixels, 16 frames per second Collect data for 3 months to get sample video images for detecting the estrus of dairy cows from their climbing behavior. Then the images are cropped with a size of 400 × 500 pixels, 300 × 300 dpi, to be used to create a classification model later in this research using the number of images. A total of 1,237 images were divided into 4 classes: the Climbing 36 images were images of cattle from the stream or other side, the Flirting 421 images were images of licking cattle, the Mating 390 images were images of cattle for breeding, and the Walking 390 images were images of cows walking.

3.2 Measuring the Efficiency of the Convolutional Neural Network

In this research, because machine learning is used, the measures used for performance are accuracy and Confusion. Matrix because it is considered an important tool for evaluating the results of predictions or predictions made from models developed in machine learning with the concept of measuring the proportion between what the researcher thought (prediction model) and what happened. What is the proportion to each other? Therefore, in summary, the confusion matrix is an important table for measuring the ability of machine learning to solve classification and prediction problems. As shown, an example using a 2 × 2 confusion matrix table is shown in Fig. 4 and the measurement details are as follows.

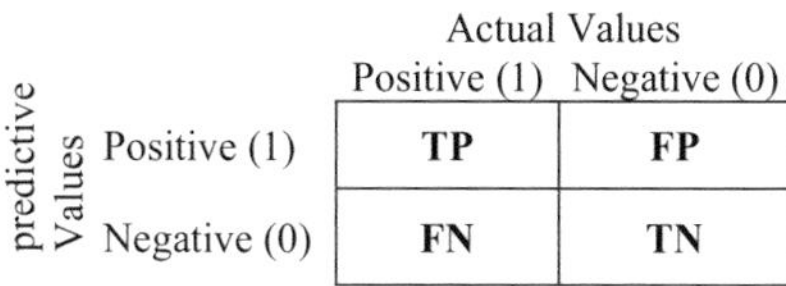

Fig. 4. The figure shows a 2 × 2 confusion matrix table.

From Fig. 4, the meaning of the table can be explained as follows. True Positive (TP) is something that the program predicts is "true" and has a value of "true". True Negative (TN) is something that the program predicts is "not true" and has a value of "not true". False Positive (FP) is something that that the program predicts is "true" but has a value of "not true". False Negative (FN) is something that the program predicts is "not true" but has a value of "true". Where TP, TN, FP, and FN in the table are represented by frequency values. Using a confusion matrix to calculate and evaluate the efficiency of model prediction, there is a calculation formula as shown in Eq. 5 as follows.

$$Accuracy = \frac{(TPs + TNs)}{TPs + TNs + FPs + FNs} \quad (5)$$

At the same time, we can calculate the prediction accuracy percentage, which is the ratio of the number of prediction accuracy to the total number of predictions. The equation is as follows in Eq. 6:

$$prediction\, accuracy = \frac{correct\, prediction}{total\, predictions} \times 100 \quad (6)$$

3.3 Configuration Hyper-parameters and Optimal Hyper-parameters with Immunity System Algorithm of Convolutional Neural Networks

To determine the parameters of the convolutional neural network, various parameters were determined using different optimizers to find the best function for predicting estrus in cows. The functions used include SGD, Adam, and RMSprop, and the cycles are set starting at 10 epochs and extending up to 100 epochs for each function to provide an appropriate value for predicting cow estrus. It was found that in the experiment of using various optimizers, the optimum value of the optimizer was Adam. Therefore, in setting the various parameters of the initial convolutional neural network before using the Artificial Immune System algorithm, it can be shown that get the sum of all parameters at 4,143,749 parameters, activation is SoftMax, the activation function is Relu, the optimizer is Adam, learning rate at 0.00000001, batch size at 128, number of cycles to run 100 epochs. After adjusting the hyperparameters from the experiment, it was found that the values obtained still had low accuracy, so to get the max accurate values. The researcher used the Immunity system algorithm to optimize the hyperparameter values to be most appropriate and vary according to the input data, as shown in the pseudocode in Fig. 5 as follows.

4 Experiments and Result

In this experiment result, the researcher presents the results obtained by using the algorithm presented in the research method above to detect estrus in dairy cows. Achieving estrus is considered the key to increasing the efficiency of the insemination of dairy cows. The duration of estrus has been studied to find the appropriate time for artificial insemination of dairy cows. In the research, the researcher prepared surveillance data of dairy cows in different positions using CCTV cameras to record moving images. Then, the researcher extracted the moving images into still images to use for predicting the total number of still images. There were 1,237 images and the researcher divided the class into 4 classes, each class containing several climbing images. There were 36 images of cows with their heads or other sides, 421 images of cows licking (flirting), then 390 images of cows climbing for mating, and 390 images of cows walking, examples of different classifications described above are shown in Fig. 6.

From the research steps mentioned above, the researcher determined the initial hyperparameter value of the convolutional neural network and adjusted the hyperparameter value to be appropriate. From the experiment, the results showed that get the sum of all parameters at 370,788 parameters, activation is SoftMax, the activation function is Relu, the optimizer is Adam, learning rate at 0.00000001, batch size at 128, number of cycles to run 100 epochs, the prediction results have an accuracy of 95.082% as shown in Fig. 7 and 8 respectively. However, the results of the experimental adjustment of the CNN parameters showed that the accuracy was not the best. Therefore, the researcher used the Artificial Immune System algorithm to adjust the parameters of the CNN, which is shown in the pseudocode in Fig. 5, to obtain the most appropriate output value that matches the input value. The fitting results show that the total number of parameters is 469,156, the activation is SoftMax, the activation function is Relu, the optimizer is Adam, the learning rate is 0.00000001, the batch size is 128, the number of cycles is 100 epochs, the accuracy value is 98.361%, comparable to Thanawat Lodkaew, et al. [13] who study estrus as a whole, i.e. mounting, sniffing and moving behavior, as estrus with an accuracy of 83.0%, but our study only classifies the specific behavior of estrus by mounting (standing heat). In addition, Nilo M. et al. [14] implemented estrus detection by mounting with an accuracy of only 50%. Eui-Ju Heo, et al. [15] use four classifications of estrus namely mounting, walking, talk wagging, and foot stamping, and then use SVM with live video stream, giving an accuracy of 91.67%. As can be seen in Figs. 9 and 10, this corresponds to the accuracy and validation graphs. The accuracy tends to increase, as can be seen in Fig. 11. This shows that the model has no problem predicting data that has never been seen well.

Algorithm Optimized Training algorithm of CNN using Immunity System Algorithm

Input:

train_x, train_y: feature and label of Training Set (Caw Mating, Walking, Flirting, Climbing)
test_x, test_y: feature and label of Test Set (Caw Mating, Walking, Flirting, Climbing)

Output:

w_{ij}^l, b_j^l: weights and bias of Convolutional and Pooling Neural Network (CPNN)
w_{jk}, b_{jk}: weights and bias of Full Connection Neural Network (FCNN)
prediction: Caw Mating, Walking, Flirting, Climbing

Required parameters:

max_time: maximum value n of nACs in every 1SP
target_error: when the current training error is less than the target error, the training is finished.
n_CPNN: the learning rate of CPNN

Initialization work:

$w_{ij}^l, b_j^l, w_{jk}, b_{jk}$: weights and scaling parameters of CNN (CPNN+FCNN) are set as random numbers.
t: t is the current simulation time, initialized as *t=1* before the training loop.
L(t): L(t) is the mean square error at simulation time *t, L(t)* is initialized as *L(1) =1>target_error.*
Artificial Immunity system algorithms
Initialization the variable of Immunity for CNN $w_{ij}^l, b_j^l, w_{jk}, b_{jk}$ then define r_{ij} matrix, number of Dedicated cell ($w_{ij}^l, b_j^l, w_{jk}, b_{jk}$)and number of shared cells, upper bound and lower bound of part in each dedicated cell

Select the first combination of parts and machines as an initial solution (S_0).
For all part and machine combinations
Generate the *popsize* (200 items) of the S_0 permutation randomly as initial population, determine the iteration value according to size of S_0
Calculate the affinity function of these antibodies (*accuracy of prediction*)
For *max gen*
Select the best n_c ($w_{ij}^l, b_j^l, w_{jk}, b_{jk}$) of the antibodies
Make (n_c^{k+1}) copies of each antibodies (clone)
Mutate all of the cloned antibodies
Calculated the affinity function value for all of the mutated antibodies
The n_c of the worse antibodies in current population are replaced by the n_c of the best mutated antibodies
End for
Store the obtained best affinity function value
End for

Begin:

Set the required parameters and complete the initialization work
While *t< max_time* and *L(t) > target_error*
for all trainingSet:
train_p (prediction of the label) is calculated according to *tarin_x* and forward calculation formula as bellow (1-9) :

$$net_{mn}^l = convolution(O^{l-1}, w^l, m, n) + b^l = \sum_{i=0}^{size^l-1}\sum_{j=0}^{size^l-1}\left(O_{m+i,n+j}^{l-1}\right).w_{i,j}^l + b^l \quad (1)$$

$$convolution(x, y) = \begin{matrix} x_{11}y_{11} + x_{12}y_{12} + x_{21}y_{21} + x_{22}y_{22} & x_{12}y_{11} + x_{13}y_{12} + x_{22}y_{21} + x_{23}y_{22} \\ x_{21}y_{11} + x_{22}y_{12} + x_{31}y_{21} + x_{32}y_{22} & x_{22}y_{11} + x_{23}y_{12} + x_{32}y_{21} + x_{33}y_{22} \end{matrix} \quad (2)$$

$$O_{mn}^l = \mathrm{F}(net_{mn}^l) = sigmoid(net_{mn}^l) = \frac{1}{1+e^{-net_{mn}^l}} \quad (3)$$

$$Y_{ij} = \mathrm{pool}(x, i, j) = \frac{\sum_{m=1}^{size^l}\sum_{n=1}^{size^l} x_{size^l \times (i-1)+m, size^1 \times (j-1)+n}^{l-1}}{size^l \times size^l} \quad (4)$$

$$\mathrm{pool}(x) = \begin{matrix} \frac{x_{11}+x_{12}+x_{21}+x_{22}}{4} & \frac{x_{13}+x_{14}+x_{23}+x_{24}}{4} \\ \frac{x_{31}+x_{32}+x_{41}+x_{42}}{4} & \frac{x_{33}+x_{34}+x_{43}+x_{44}}{4} \end{matrix} \quad (5)$$

$$O_i^{-2} = O_{mn}^{-3}, m = int\left(\frac{i}{size^{-2}}\right) + 1, n = i - size^{-2} \times (m-1) \quad (6)$$

Fig. 5. The figure shows the pseudocode of the CNN optimized algorithm for the immune system.

$$net_j^{-1} = \sum_{i=1}^{size^{-2}} \left(O_i^{-2} \,.\, w_{ij}^{-1} + b^{-1}\right), j = 1,2, \ldots, size^{-1} \tag{7}$$

$$O_j^{-1} = \mathrm{F}\left(net_j^{-1}\right) = \mathrm{sigmoid}\left(net_j^{-1}\right) = \frac{1}{1+e^{-net_i^{-1}}} \tag{8}$$

$$\hat{y}_n = o^{-1} \tag{9}$$

end for

L(t) is re-calculated as ***L(t)*** $= \frac{1}{2}\sum_{n=1}^{N}\left(train_{p(n)} train_{y(n)}\right)^2$, N is the total number of training Set

$\Delta w^l, \Delta b^l, \Delta w_{ij}^{-1}, \Delta b^{-1}$ are updated according to the formula (20-23*)* as bellow:

$$\Delta w^l = \frac{\partial L}{\partial w^l} = \frac{\partial L}{\partial net^l} \times \frac{\partial net^l}{\partial w^l} = \delta^l . O^{l-1} \tag{20}$$

$$\Delta b^l = \frac{\partial L}{\partial b^l} = \frac{\partial L}{\partial net^l} \times \frac{\partial net^l}{\partial b^l} = \delta^l \tag{21}$$

$$\Delta w_{ij}^{-1} = \frac{\partial L}{\partial w_{ij}^{-1}} = \frac{\partial L}{\partial net_j^{-1}} \times \frac{\partial net_j^{-1}}{\partial w_{ij}^{-1}} = \delta_j^{-1} . O_i^{-2} \tag{22}$$

$$\Delta b^{-1} = \frac{\partial L}{\partial b_j^{-1}} = \frac{\partial L}{\partial net_j^{-1}} \times \frac{\partial net_j^{-1}}{\partial b_j^{-1}} = \frac{1}{size^l} \sum_{j=1}^{size^{-1}} \delta_j^{-1} \tag{23}$$

$w^l(t), b^l(t), w_{ij}^{-1}(t), b_j^{-1}(t)$ are adjusted according to the formula (24-27) as bellow:

$$w^l(t+1) = w^l(t) - \eta\_\mathrm{CPNN} \times \Delta w^l \tag{24}$$

$$b^l(t+1) = b^l(t) - \eta\_\mathrm{CPNN} \times \Delta b^l \tag{25}$$

$$w_{ij}^{-1}(t+1) = w_{ij}^{-1}(t) - \eta\_\mathrm{CPNN} \times \Delta w_{ij}^{-1} \tag{26}$$

$$b^{-1}(t) = b^{-1}(t) - \eta\_\mathrm{CPNN} \times \Delta b_j^{-1} \tag{27}$$

t++

end while

End

Fig. 5. *(continued)*

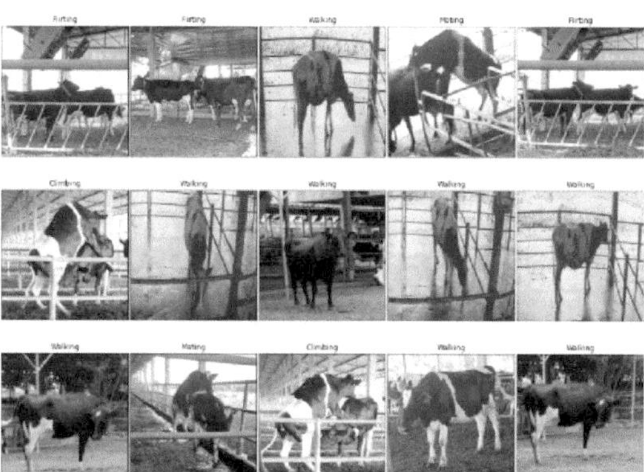

Fig. 6. The figure shows an example image of a dairy cow with 4 movements used for the prediction.

	precision	recall	F1-score	support
0	1.00	1.00	1.00	1
1	0.91	1.00	0.95	20
2	1.00	0.94	0.97	18
3	0.95	0.91	0.93	22
accuracy			0.95	61
macro avg	0.97	0.96	0.96	61
weighted avg	0.95	0.95	0.95	61

Confusion Matrix: [[1 0 0 0]
[1 20 0 0]
[0 0 17 1]
[0 2 0 20]]

Accuracy: 0.95082

Fig. 7. The confusion matrix and accuracy of CNN by testing cow estrus dataset.

Fig. 8. The image shows the results obtained from the prediction using the CNN, the accuracy is 95.082%.

	precision	recall	F1-score	support
0	1.00	1.00	1.00	1
1	1.00	0.95	0.97	20
2	1.00	1.00	1.00	18
3	0.96	1.00	0.98	22
accuracy			0.89	61
macro avg	0.99	0.99	0.99	61
weighted avg	0.98	0.98	0.98	61

Confusion Matrix: [[1 0 0 0]
[1 19 0 1]
[0 0 18 0]
[1 0 0 22]]

Accuracy: 0.98361

Fig. 9. The confusion matrix and accuracy of CNN optimized with Immunity System by testing cow estrus dataset.

Fig. 10. The image shows the results obtained from the prediction using the CNN optimized with Immunity System, the accuracy is 98.361%.

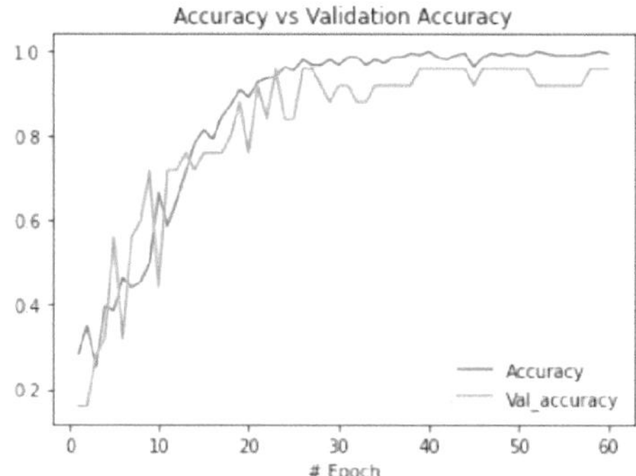

Fig. 11. The accuracy tends to increase between accuracy and validation accuracy.

5 Conclusion

Due to the origin and importance of the problem in monitoring estrus in dairy cows, which affects the production of drinking milk and other industries that use milk as a raw material, various research studies on estrus in dairy cows have shown that a significant percentage of estrus occurs at night, making monitoring difficult. Therefore, research has found the use of various tools. They help to detect and monitor estrus. However, this comes at quite a high cost (as explained in the introduction and problem section). In this study, a method is used that does not require a high investment, namely CCTV video monitoring. Capturing the behavior of dairy cows and using developed algorithms to detect estrus may not be very convenient because the images must first be converted to still images and then serve as input for the algorithm developed for this research (Figs. 12 and 13).

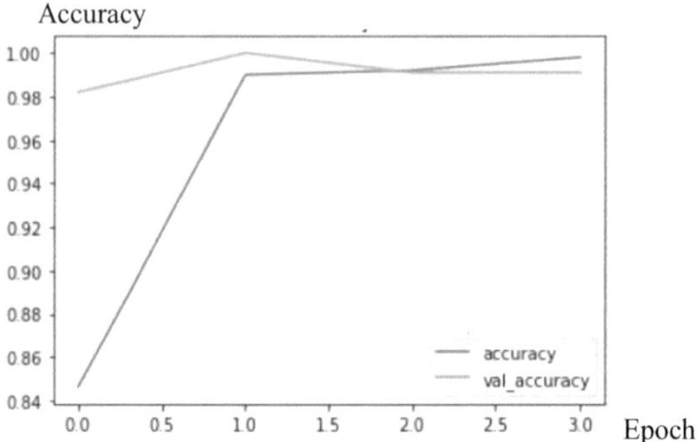

Fig. 12. The accuracy graph is shown between the accuracy and validation accuracy of CNN optimized by AIS.

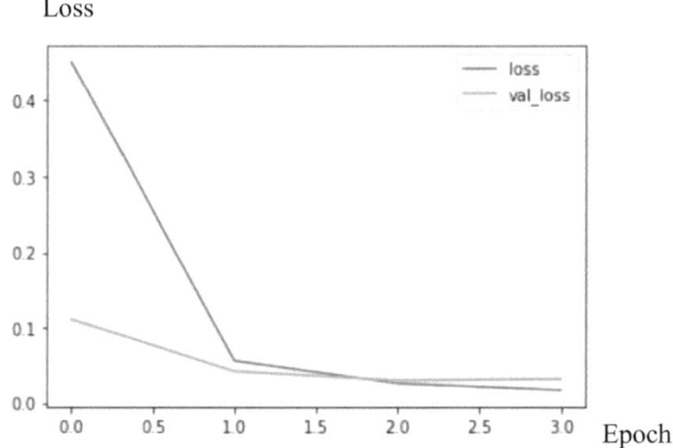

Fig. 13. The loss graph is shown between the loss and value loss of CNN optimized by AIS.

At a time when artificial intelligence is being developed and used for various tasks, the use of artificial neural networks in work related to the use of artificial intelligence naturally plays a role, which is widely recognized by many. In various research studies, it has been found that most researchers use convolutional neural networks when predicting or classifying image data because they provide the best output values. However, there are problems that researchers encounter. The most important thing is to tune the parameters of this convolutional neural network to the input data to get the best output values. If this is the case, the researcher in this research needs to develop a method to adjust the parameters of this type of artificial neural network by using the artificial immunity system algorithm to adjust the number of layers of the network and the number of hidden node layers and the bias value by comparing the output value, which is the accuracy value before using the mentioned algorithm to adjust and using the mentioned algorithm to adjust the parameters.

The results of the experiment show that the accuracy obtained before using the above algorithm to fit the parameters was 95.082%, while after using the Artificial Immunity System algorithm to fit the parameters, an accuracy value of 98.361% was obtained, which can be seen in the experimental results from using the confusion matrix, as shown in Figs. 7 and 9, respectively 9, and when looking at the loss graph and the value graph, the loss is found to be well fitting, which is the goal of training the model. A good fitting graph indicates that the model has learned well. Researchers can use the model to predict data that they have never seen before, or it is referred to as a model that can generalize to new data (with less generalization error). However, this research found that the actual application of heat detection in dairy cows needs further development. For example, heat detection in dairy cows requires timely detection and warning to dairy farmers so that they can breed in time, because if they cannot breed in time, they may have to wait for a new heat round in the next 14 days. Therefore, the researchers will further develop the detection of cow estrus using CCTV cameras to detect movements in real time.

References

1. NATDA newsletter. Biotech articles develop technology to promote the dairy farming industry. NSTDA 12 (2016)
2. DDP Team. The situation of world and Thai dairy cattle in 2019. Bureau of Agricultural Economic Research (2019)
3. Suneerat, I.: Reproductive management in dairy cattle. Dairy Newsl. **19**(10), 1–5 (2018)
4. Department of Livestock Development, Ministry of Agriculture and Cooperatives. Beef cattle health care. In the Livestock Development and Technology Transfer Office, Department of Livestock Development, Beef Cattle Health Care, 1(1–52). Agricultural Cooperatives Association of Thailand Printing Co., Ltd. Bureau of Animal Disease Control, Prevention and Treatment (2006)
5. Serap, G., Nazan, K.: The sensor technologies for more efficient cow reproduction systems. MOJ Ecol. Environ. Sci. **4**(3), 128–131 (2019)
6. Thi, T.Z., et al.: Estrus detection for dairy cow using a laser range sensor. In: Third International Conference on Computing Measurement Control and Sensor Network, pp. 162–165. IEEE Press (2016). https://doi.org/10.1109/CMCSN.2016.50
7. Ali, A.S., et al.: Estrus detection in a dairy herd using an electronic nose by direct sampling on the perineal region. Veterinary Sci. **9**(688), 1–12 (2022). https://doi.org/10.3390/vetsci9120688
8. Shogo, H., et al.: Estrus detection using background image subtraction technique in tie-stalled cows. Animals **11**(6), 1795, 1–8 (2021). https://doi.org/10.3390/ani11061795
9. Zhen, W., et al.: A non-contact cow estrus monitoring method based on the thermal infrared images of cows. Agriculture **13**(2), 385, 1–19 (2023). https://doi.org/10.3390/agriculture13020385
10. Juan, R.A., et al.: Estimating body condition score in dairy cows from depth images using convolutional neural networks, transfer learning, and model ensembling techniques. Agriculture **9**(2), 90, 1–18 (2019). https://doi.org/10.3390/agronomy9020090
11. Alexey R., et al.: Live weight prediction of cattle based on deep regression of RGB-D images. Agriculture **12**(11), 1794, 1–17 (2022). https://doi.org/10.3390/agriculture12111794
12. Daniel, N.: What are CNNs (convolutional neural networks)?. Unite AI (2020). https://www.unite.ai/th/Whatisanartificialneuralnetwork/
13. Thanawat, L., Kitsuchart, P., Loo, C.K.: CowXNet: an automated cow estrus detection system. Expert Syst. Appl. **211**(118550) (2023). https://doi.org/10.1016/j.eswa.2022.118550

14. Nilo, M.A., et al.: Automated estrus detection for dairy cattle through neural networks and bounding box corner analysis. Int. J. Adv. Comput. Sci. Appl. **11** (2020). https://api.semanticscholar.org/CorpusID:223039807
15. Eui, J.H., Sung, J.A., Kang, S.C.: Real-time cattle action recognition for estrus detection. Ksii Trans. Internet Inf. Syst. **13**(4), 2148–2161 (2019)

Sequential Data Approach for Rate of Penetration Prediction Using Machine Learning Models: A Case Study the Offshore Volve Oil Field, North Sea, Norway

Yanadade Pakawatthapana(✉) and Subhorn Khonthapagdee(✉)

Department of Computer Science, Srinakharinwirot University, Bangkok 10110, Thailand
{yanadade.pak,subhorn}@g.swu.ac.th

Abstract. In drilling operations, the Rate of Penetration directly impacts drilling efficiency. Accurately predicting this rate is crucial as it affects the cost-effectiveness of the process. In this study, ensemble machine learning models such as Random Forest, Gradient Boosting, Extreme Gradient Boosting, Adaptive Boosting, and K-Nearest Neighbor were used to predict ROP using a Volve dataset from the Norwegian North Sea provided by Equinor Company. To simulate real-world scenarios, especially in cases with limited amounts of data, we utilized sequential data or depth-dependent data to trained and tested various machine learning models using sequential split such as continuous learning and sliding window split techniques. The model's performances were evaluated by calculating mean absolute error. The results of this study showed that the ensemble machine learning models, especially Extreme Gradient Boosting with sliding window split technique outperformed in predicting the Rate of Penetration. Moreover, the various sizes of the sliding window were studied to find the optimal size.

Keywords: Rate of Penetration · Drilling Operations · Machine Learning Models · Ensemble Models · Continuous Learning · Sliding Window

1 Introduction

The oil and gas industry heavily relies on optimized drilling operations to enhance the efficiency and cost-effectiveness of hydrocarbon extraction. A crucial factor that directly impacts drilling efficiency is the Rate of Penetration (ROP), which measures the speed at which a drilling bit advances into subsurface formations. Predicting ROP is essential for efficient drilling operations, but it can be challenging due to its dependence on various factors, such as drilling fluid properties, parameters, and formation characteristics. Several factors influence the rate of penetration during drilling operations including weight on bit, rotary speed, drilling fluid properties such as viscosity and density, bit type and design, and formation properties. Traditional methods for ROP prediction often rely on empirical models and expert knowledge, which may lack the adaptability required for the dynamic and complex nature of drilling environments. Recently, data-driven techniques

P. Meesad et al. (Eds.): IC2IT 2024, LNNS 973, pp. 121–130, 2024.
https://doi.org/10.1007/978-3-031-58561-6_12

like machine learning, particularly regression models, have shown promising results in predicting drilling-related parameters. In recent years, data-driven approaches, such as machine learning (ML) models, have gained attention in the oil and gas industry for their potential to accurately predict the rate of penetration during drilling operations. These approaches leverage large datasets containing various drilling parameters, such as weight on bit, rotary speed, drilling fluid properties, bit type, and formation characteristics, to develop predictive models. Usually, Machine learning models rely on training and testing with historical drilling data to develop accurate predictive models for the rate of penetration. However, in drilling an exploration well, the information gathered from subsurface formations is often limited and data starts to become available once the drilling operation begins. To mimic real-time ROP prediction, this study aims to compare the prediction performance of Random Forest (RF), Gradient Boosting (GB), Extreme gradient boosting (XGB), Adaptive boosting (AdaB), and K-Nearest Neighbor (KNN) using two different techniques for training and testing on depth-dependent drilling datasets.

2 Background and Related Work

Traditional methods for predicting ROP typically rely on physics-based models utilizing formulas and mathematical functions. Speer's model [1] is based on a mathematical approach that considers key drilling parameters, including weight-on-bit, rotary speed, and hydraulic power. Similarly, Bingham's model [2] also incorporates formation type. Additionally, [3] and [4] further considered rock strength in their models. These traditional models, while useful, often require the evaluation of various constants based on field results or experimental methods, leading to time consumption and limited generalization for the entire well [5]. Recent developments in machine learning and big data infrastructure have allowed for the development of data-driven models that can predict ROP more accurately and efficiently. The study by [6] showed that Random forests achieve higher accuracy and goodness of fit (R^2) compared to traditional physics-based models. [7] conducted another study that concluded ML algorithms, including random forest, support vector machine, and neural network, substantially outperform traditional ROP models. Therefore, machine learning models have emerged as a promising approach for ROP prediction in drilling [8]. Ensemble machine learning combines multiple base models to form a stronger predictive model, leveraging the strengths of individual models to improve overall performance. These ensemble such as Random Forest, gradient boosting, adaptive boosting models have shown high accuracy and robustness in predicting the rate of penetration [6, 7, 9]. K-Nearest Neighbor is a non-parametric classification and regression algorithm that can be effectively used for predicting the Rate of Penetration. Unlike traditional parametric models, KNN does not make any assumptions about the distribution of the data and instead directly uses the available dataset for making predictions. The use of KNN for ROP prediction has been supported by several studies, including the work by [10] and [11].

When developing accurate predictive models, the train-test split in machine learning is a crucial step. This involves dividing the available dataset into two subsets: the training set and the testing set. The training set is used to train the machine learning models, while the testing set is used to evaluate their performance. The conventional approach for data splitting is to do so randomly; however, [6] and [12] suggested treating the dataset as sequential data to simulate drilling processes. Sequential splitting preserves depth dependencies and patterns in the drilling process. [12] considered a scenario known as "continuous learning," which involves training and testing the model every 30 m. This data split is similar to a forward-chaining split or a rolling-origin approach in time series analysis, where the training and testing sets are sequentially partitioned to ensure that the model is trained on past data and tested on future data [13].

Our contributions include an extension of the research by [12], with a focus on continuous learning scenarios. We analyze and compare the performance of different machine learning algorithms using two sequential split approaches: continuous learning (forward-chaining split) and sliding window splits. Furthermore, we explore how varying window sizes in the sliding window approach affect the predictive accuracy of ROP models.

3 Data Preparation

3.1 Data Source

In this research study, real offshore field data was operated between 2008 and 2016 by Equinor Company, Norway [12]. The dataset from the Volve oil field, located on the Norwegian Continental Shelf (NCS) in the central part of the North Sea, approximately 200 km west of Stavanger and 5 km north of Sleipner East, was made open source in 2018.

3.2 Data Acquisition

The Volve field dataset can be accessed through the Equinor Open Data License, which can be accessed on [14]. Our study focused on drilling of an exploration well 15/9-F-15, StatoilHydro (Norway-StatoilHydro-15_47_9-F-15). The drilling operation data involves two sections: the 17.5″ section and the 8.5″ section. The data was stored in Well-site Information Transfer Standard Markup Language (WITSML) and was imported and analyzed in Python. The process from getting data and import into Python can be found in [15] and [16]. The raw data consists of 112,714 rows and 189 columns with the depth starting from 1178.509 m to 4065.346 m.

3.3 Data Cleaning and Pre-Processing

In this subsection, we provide detail on how to pre-process and clean the dataset. We removed data from the starting point up to 1200 m after discovering that the logged data was not recorded. The drilling operation for the 17.5″ section reached a depth of 2591.141 m, followed by pre-drilling for the 8.5″ sections starting from 2464.156 m, which led to unnecessary duplication that needs to be removed. The column indicating the type of section was renamed as Bit Diameter. The missing values were filled using Forward Fill and Backward Fill methods. The measured depth was rounded to the nearest integer, and redundant depth entries were removed, keeping only the initial value as a representation for that depth. A new column Total Flow Area was also added. Finally, the dataset consists of 2,701 rows and includes a selection of 10 columns based on expert knowledge.

3.4 Feature Selection

In the context of ROP prediction, the input variables typically consist of drilling parameters such as weight on bit, rotary speed, drilling fluid properties, bit type, and formation characteristics [7, 17, 18]. These features are crucial as they directly influence the drilling process and affect the rate of penetration. For this study, our selected features for machine learning models include Rate of Penetration (ROP), Measure Depth (MD), Weight on Bit (WOB), Rotary Speed (RPM), Surface Torque (T), Standpipe Pressure (SPP), Mud Flow In (Q), Mud Weight Out (MW_{out}), Bit Diameter (D_{BIT}) and Total Flow Area (TFA). The features and details of drilling techniques were shown in Table 1.

Table 1. The features and details of drilling techniques.

Feature	Definition	Unit
Rate of Penetration (ROP)	A measure of the speed at which the bit drill into formation	Meter/hour (m/hr)
Measure Depth (MD)	The length of the drilled borehole	Meter (m)
Weight on Bit (WOB)	The amount of downward force placed on the Bit	Kilo-kilogram force (kkgf)
Rotary Speed (RPM)	The speed at which the drill string and bit rotate during the drilling process	Revolution per minute (rpm)
Surface Torque (T)	The amount of rotational force or torque applied at the surface of the wellbore	Kilonewtons. Meter (kN.m)
Standpipe Pressure (SPP)	The pressure of the drilling fluid within the standpipe	Kilopascal (kPa)
Mud Flow In (Q)	Circulation of drilling mud into the wellbore during the drilling process	Liter/minute (L/min)
Mud Weight Out (MW_{out})	The process of measuring and adjusting the density or weight of the drilling fluid as it returns through the flowline	Grams / cubic centimeter (g/cm^3)
Bit Diameter (D_{BIT})	The diameter of the drill bit	Inch
Total Flow Area (TFA)	The set of nozzle areas which fluid flow through a bit	Inch^2

4 Exploratory Data Analysis (EDA)

In order to gain insights into the dataset and understand the relationships between variables, an exploratory data analysis was conducted. From a dataset of 2,701 rows and 10 columns, outliers of Rate of Penetration (ROP) greater than 400 were removed, resulting in a remaining dataset of 2,664 rows and 10 columns. At a depth between 2,592–2,604, an unusually high Weight on Bit was detected as a result of changes in the section. The abnormal values were removed and the data was corrected using linear interpolation. The Rate of Penetration in Fig. 1.(a) starts off at a high level, maintaining this between the depths of 1,200 and 1,306 during initial drilling. Afterward, the rate of penetration declines and stabilizes around depth 1,320. In the 17.5″ section, the penetration rate ranges from 5 to 25, while in the 8.5″ section it varies from 20 to 40, with occasional rapid increases occurring at specific intervals such as depth 3,151–3,177 due to changes in rock formation presence. Weight on Bit (WOB) and rotary speed (RPM) were among the drilling parameters that showed a strong correlation with the rate of penetration. In Fig. 1.(b), it can be observed that the WOB initially ranges between 0 and 5 in the 17.5″ section during the early drilling stages. Subsequently, there is an increase to a range of 5–11 (depth between 2,213–2,591). As for the 8.5″ section, WOB increases until it reaches a value of 13, after which it is reduced (depth between 2,951–3,517). The average WOB during this range is found to be around 6. Following this decrease, WOB increases again with values ranging between 7 and19. Rotary Speed (RPM) in Fig. 1.(c) - In the 17.5″ section, RPM gradually increases until reaching depth 1,480. After that point, RPM stabilizes at around 130 rpm. As for the 8.5″ section, RPM is approximately 140 rpm.

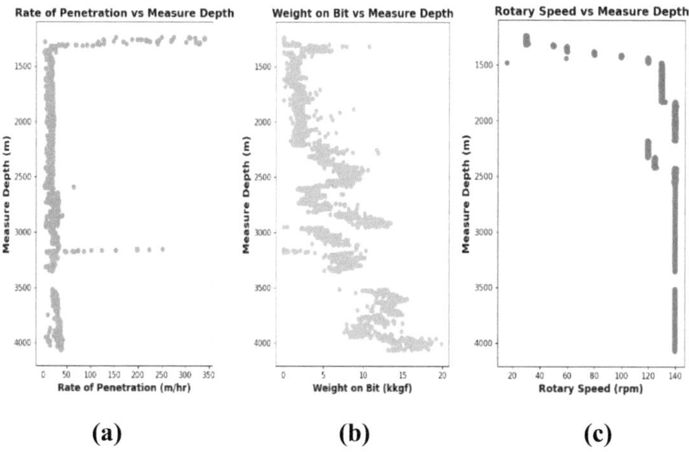

Fig. 1. Explaining features compared to Measure Depth. (a). Rate of Penetration (b). Weight on Bit (c). Rotary Speed.

5 Model Setup and Training

5.1 Training Machine Learning Model

In this study, we treat the dataset as sequential data or depth-dependent data to simulate drilling processes. Therefore, models were trained and tested using sequential split approaches. Two techniques were employed: **continuous learning** (forward-chaining split) and **sliding window split**. Both techniques require the data to be split up to a certain depth, which is referred to as the "increment." This increment represents the interval of depth at which the data is split in each iteration. The increments of 5, 10, 15, 20, 50, 100, 200 and 500 were chosen for the experiments to evaluate their respective effects on the accuracy of the machine learning model.

Figure 2 illustrates the distinction between continuous learning and sliding window split. Initially, both techniques have identical sizes for the training set and testing set. However, in second iterations, the continuous learning approach integrates new data with the existing training set, whereas the sliding window split approach slides the window of training and testing data forward while keeping the size of training and testing sets equal to the increment.

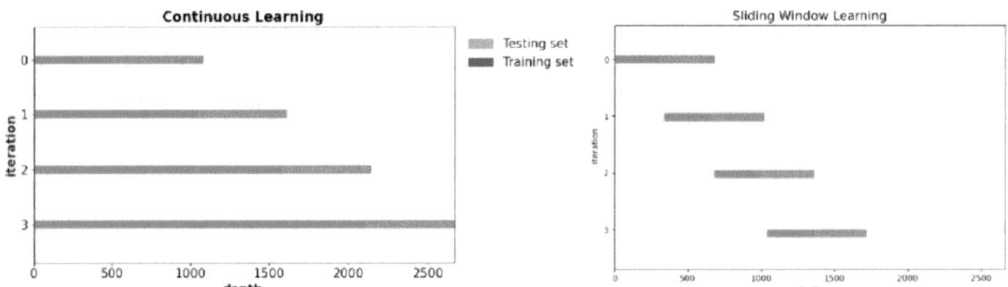

Fig. 2. Illustrates the distinction between continuous learning and sliding window split.

5.2 Model Setup and Evaluation

Five models were used in this study. They are Random Forest (RF), Gradient Boosting (GB), Extreme gradient boosting (XGB), Adaptive boosting (AdaB), and K-Nearest Neighbor (KNN). All experiments were conducted using the scikit-learn library in Python with default parameters. For sequential split approaches, we adapted the GitHub "continuous.py" code [19] to suit our needs. For continuous learning, we simply adjusted the number of increments. Conversely, for sliding window splits, we modified the while loop to ensure the next iteration's training set originated from the current testing set. We assessed all models using mean absolute error (MAE), as defined in [20]. Unlike other metrics such as Mean Square Error or Root Mean Square Error, MAE provides a straightforward measure of the average magnitude of errors in the predictions without exaggerating the contribution of large errors. It is worth noting that we also set the random seed of numpy and random state in scikit-learn to ensure consistency and reproducibility of the results [21].

6 Results

In each iteration, the model was trained and tested using the specified split technique (continuous learning or sliding window) with a particular increment value. Only the MAE score for the test dataset was recorded. The average MAE from all iterations was calculated for each model and split technique to determine overall performance. Figure 3 illustrates the comparison of MAE of test data for different models across various increments.

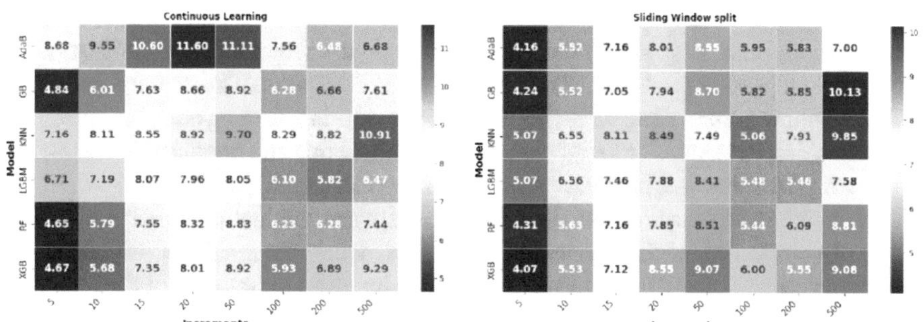

Fig. 3. The average MAE for different models across 8 increments (a) continuous learning (b) sliding window split.

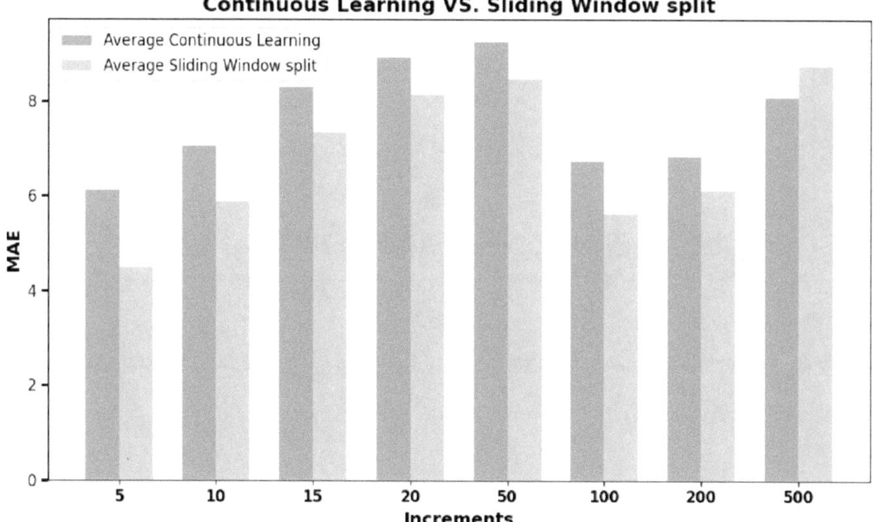

Fig. 4. The average MAE of continuous learning vs sliding window split.

To see the overall performance of each split technique, the average MAE from all model was calculated for each increment and presented in Fig. 4. The sliding window split performed better in terms of MAE for the test data compared to continuous learning, except for an increment of 500 where the continuous learning approach showed slightly lower MAE. It showed that as the increment increased, the error also increased until an increment of 50, then decreased again at 100 before rising once more. In terms of model performance, ensemble models like XGB, RF, and GB showed lower MAE with smaller increments such as 5 and 10, then again with larger increments of 100 and 200. However, LGBM and AdaB exhibited lower MAE with higher increments such as 100, 200, and 500.

Figure 3 shows that the XGB model with a 5-increment sliding window split achieved the lowest Mean Absolute Error (MAE) on the test data. Its goodness of fit (R^2) was 0.88 [22]. To visualize this performance, Fig. 5 compares the true and predicted ROP values using a scatter plot. Blue dots represent true ROP values from the entire dataset, while yellow dots represent predicted ROP values from each iteration. Due to the 5-increment sliding window nature, there were 533 iterations of training, testing, and predicting, and only the predicted values were recorded. We observed that the predicted value closely aligns with the true value, but there were certain depths that deviated from the prediction. The residuals were calculated as the difference between the true and predicted values. The dashed line in Fig. 5 indicates the interval at which the residuals fall outside an acceptable range, set at 3 times the standard deviation from the average of the residuals.

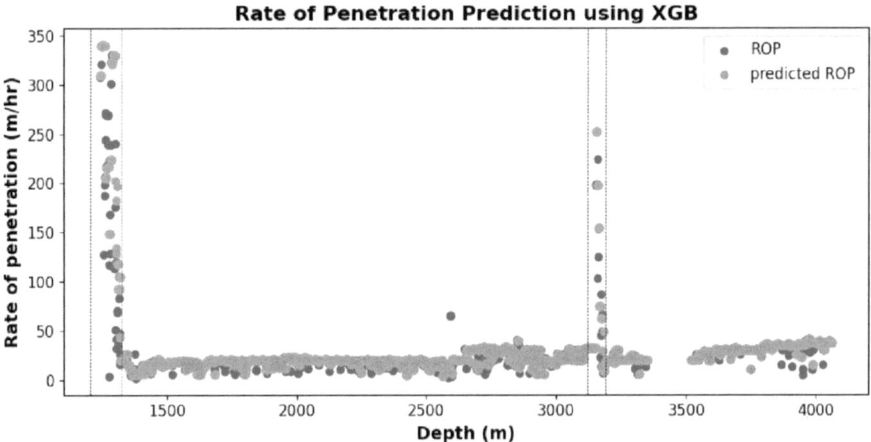

Fig. 5. The comparison between the true value of ROP and predicted value of ROP using XGB.

7 Conclusion

This study utilized sequential split approaches and machine learning models for simulating drilling processes and predicting ROP based on depth-dependent data. The findings suggest that for this dataset the sliding window split technique with a small increment size and ensemble models yielded the best performance in predicting the rate of penetration. Using a small increment is advantageous when we have limited training data, but it requires many iterations to train the model. Once enough data is available, increasing the increment size can reduce the number of iterations needed for this dataset. The formation properties are key factors to indicate size of the increment. A smaller increment signifies that the data remains within the same formation. Many studies have already shown that ensemble models such as XGBoost, Random Forest, Gradient Boosting, and KNN tend to perform well in predicting ROP. This study emphasizes the importance of considering depth-dependent data and incorporating it into prediction models to simulate real-time drilling operations for exploration wells. For future work, it would be interesting to further explore the cross validation method for hyperparameter tuning in each training iteration.

Acknowledgements. We would like to thank the instructors from the Department of Computer Science at Srinakharinwirot University, Thailand. In addition, this research was funded by the Graduate School of Srinakharinwirot University to support the publication.

References

1. Speer, J.W.: A method for determining optimum drilling techniques. In: SPE Upper Gulf Coast Drilling and Production Conference, SPE-1242, SPE (1959)
2. Bingham, M.G.: A New Approach to Interpreting Rock Drillability (1964)
3. Hareland, G., Rampersad, P.R.: Drag-bit model including wear. In: SPE Latin America and Caribbean Petroleum Engineering Conference, SPE-26957 (1994)
4. Motahhari, H.R., Hareland, G., James, J.A.: Improved drilling efficiency technique using integrated PDM and PDC bit parameters. J. Can. Pet. Technol. **49**(10), 45–52 (2010)
5. Chandrasekaran, S., Kumar, G.S.: Drilling efficiency improvement and rate of penetration optimization by machine learning and data analytics. Int. J. Math. Eng. Manag. Sci. **5**(3), 381–394 (2020)
6. Hegde, C., Daigle, H., Millwater, H., Gray, K.: Analysis of rate of penetration (ROP) prediction in drilling using physics-based and data-driven models. J. Petrol. Sci. Eng. **159**, 295–306 (2017)
7. Soares, C., Gray, K.: Real-time predictive capabilities of analytical and machine learning rate of penetration (ROP) models. J. Petrol. Sci. Eng. **172**, 934–959 (2019)
8. Ren, Y., Lu, B., Zheng, S., Bai, K., Cheng, L., Yan, H., Wang, G.: Research on the Rate of Penetration Prediction Method Based on Stacking Ensemble Learning. Geofluids 2023 (2023)
9. Al-Sahlanee, D.T., Allawi, R.H., Al-Mudhafar, W.J., Yao, C.: Ensemble Machine Learning for Data-Driven Predictive Analyt-ics of Drilling Rate of Penetration (ROP) Modeling: A Case Study in a Southern Iraqi Oil Field. SPE Western Regional Meeting, SPE-213043-MS, D021S004R007, SPE (2023)
10. El-Sayed, Y., Salem, A.M., El-Rammah, S.: Rate of penetration prediction in drilling operation in oil and gas wells by k-nearest neighbors and multi-layer perceptron algorithms. J. Min. Environ. (JME) **14**(3), 755–770 (2023)
11. Olafadehan, O.A., Ahaotu, I.D.: Comparative analysis of machine learning algorithms in predicting rate of penetration during drilling. J. Pet. Chem. Eng. **1**(1), 32–47 (2023). https://urfpublishers.com/journal/petrochemical-engineering
12. Tunkiel, A.T., Sui, D., Wiktorski, T.: Reference dataset for rate of penetration benchmarking. J. Petrol. Sci. Eng. **196**, 108069 (2021)
13. Tashman, L.J.: Out-of-sample tests of forecasting accuracy: an analysis and review. Int. J. Forecast. **16**(4), 437–450 (2000)

14. Equinor.: Volve field data set. https://www.equinor.com/energy/volve-data-sharing. Accessed 14 Jan 2024
15. Bansal, A.: DiscoverVolve: Understanding the WITSML format in the Volve Oil-field dataset. https://discovervolve.com/2021/01/06/understanding-the-witsml-format-in-the-volve-oilfield-dataset/. Accessed 14 Jan 2024
16. Postanes, M.: Interactive-Well-Trajectory-Plot Github. https://github.com/maribickpostanes/Interactive-Well-Trajectory-Plot. Accessed 14 Jan 2024
17. Ashrafi, S.B., Anemangely, M., Sabah, M., Ameri, M.J.: Application of hybrid artificial neural networks for predicting rate of penetration (ROP): a case study from Marun oil field. J. Petrol. Sci. Eng. **175**, 604–623 (2019)
18. Gan, C., et al.: Prediction of drilling rate of penetration (ROP) using hybrid support vector regression: a case study on the Shennongjia area, Central China. J. Petrol. Sci. Eng. **181**, 106200 (2019)
19. Tunkiel, A.T.: USOP Github. https://github.com/AndrzejTunkiel/USROP. Accessed 14 Jan 2024
20. Wang, W., Lu, Y.: Analysis of the mean absolute error (MAE) and the root mean square error (RMSE) in assessing rounding model. In: IOP Conference Series: Materials Science and Engineering, vol. 324, p. 012049 (2018)
21. Arenas R., G.: Reproducibility. https://sklearn-genetic-opt.readthedocs.io/en/stable/tutorials/reproducibility.html. Accessed 14 Jan 2024
22. Scikit-learn developers (BSD License): Metrics and scoring: quantifying the quality of predictions. https://scikit-learn.org/stable/modules/model_evaluation.html#r2-score. Accessed 14 Jan 2024

3D Liver Segmentation from CT-Scan Images

Nateepat Sutiratanapong(✉) and Tanasai Sucontphunt

National Institute of Development Administration, 148 Seri Thai Road, Khlong Chan, Bang Kapi, Bangkok 10240, Thailand
nateepat.sut@stu.nida.ac.th

Abstract. Segmenting the liver from volumetric radiographs is a time-intensive task. Machine learning can be utilized to automate this process when precision is not the primary goal. The development of machine learning models is mainly constrained by GPU memory. To preserve spatial localization in segmentation, processing the entire volumetric image at once is ideal, but this approach is memory-intensive. Sophisticated models can extract more detailed features, but may require dividing the volumetric image into smaller segments to fit within memory constraints. Conversely, simpler models, while less memory-intensive, might not capture all necessary features for accurate segmentation. These three requirements – precision, memory usage, and feature extraction – have their own trade-offs and thus become the focus of this research. In this paper, a simple U-Net model is employed to process the entire volumetric image, thereby preserving spatial localization. This non-complex model is enhanced by a special technique, Organ-Centric Resampling, introduced for data preprocessing. Independent Random Fields Deformation is also applied to increase the variety of the generated image.

Keywords: Organ-Centric Resampling · Organ Segmentation · Volumetric Segmentation

1 Introduction

Segmentation requires a physician's specialty to extract meaningful data out of a 3D radiograph, on a slice basis. The task is typically a prerequisite prior to surgical planning which may require a hand-drawn areas of interest, to facilitate an understanding of the patient's anatomical structure. Nonetheless, a considerable amount of time may be ineffectively expended in conducting this tedious task. This research aims to provide automated segmentation that allows radiologists and physicians to spend their extra time to treat more serious patients.

A volumetric radiograph is popular for surgical planning or creating dose plans used for treating cancer. When a volumetric scan is acquired, it can be depicted as a dense cube with intensity in each of its voxels, varying in Hounsfield Unit (HU). In other words, segmentation is a step used for extracting meaningful data from a dense cube, which requires human processing. Each volumetric radiograph is comprised of many

P. Meesad et al. (Eds.): IC2IT 2024, LNNS 973, pp. 131–140, 2024.
https://doi.org/10.1007/978-3-031-58561-6_13

slices of two-dimensional spatial data, and each slice needs to be hand-drawn, making this a laborious task.

Machine learning is commonly used for 3D automatic segmentation tasks because there are many parameters to consider in processing the segmentation, where traditional methods might not be as effective. Machine learning requires a model, and once the training session trains it with data to extract meaningful features, a trained model file is created. Convolution is a mathematical operation that applies filters to the image to extract features such as textures or edges, and is popular in performing segmentation tasks. The traditional machine learning technique involves stacking convolutional layers which extract features as they proceed to deeper ones, but the downside is gradient descent, where the learned performances are too insignificant and wouldn't contribute to model optimization. U-Net, which has gained popularity for its skip connections, captures localized features as the data proceeds through deeper layers like its predecessor, but it allows skipping connections to preserve detailed segmentation, solving the gradient descent issue. Many sophisticated models are being developed every day, but it's also possible that the more advanced model might also require more resources. This has led many developers to process volumetric data on a slice-by-slice basis instead, after which these slices are stacked to form a complete 3D object. Some separate the volumetric data into patches of small dice shapes, which are then joined together to form the full 3D object as a final product. While this approach allows for the extraction and formulation of more data, it might also affect the ability of complex models to localize features. This paper will focus on a model that can preserve spatial localization.

Augmentation is the technique to increase the number of images that are being supplied to the model. Since the model in the training session learns from the limited input images provided, a variety of images is required, or the model would only remember the features of the input and would not perform its task as expected once untrained data are used as the input. This phenomenon is called overfitting. To prevent this, several augmentations have been developed. Popular augmentation techniques include rotating, flipping, zooming, and cropping, etc. There are many tools, but not all are suitable for medical segmentation tasks. Rotation will be used as it can occur naturally. Flipping will be avoided as human organs will never be flipped. Zooming is inadequate for this project as we have implemented another technique to crop the image; such an operation would cut the liver.

To focus on the importance of volumetric spatial aspects when the model is extracting features, the entire 3D volume should be supplied to the model without splitting. The process is memory-intensive, and thus the model cannot be too complex to ensure compatibility with the limited memory in the GPU. However, a tradeoff exists: too simple models might not extract adequate features. This issue becomes the focus of this research. We can enhance a simpler model's performance with data preprocessing and augmentation.

Furthermore, a Hybrid loss, comprising categorical cross entropy and Dice loss, will be utilized to further facilitate gradient optimization. IOU metrics will be used to monitor an adequate time to stop the training to prevent the model from becoming overfit. These losses and metrics support volumetric input, and the paper utilizes modules from Segmentation Models 3D [1] and Classification models 3D [2]. The project is developed using the Keras [3] and TensorFlow [4] framework. For preprocessing tasks and augmentation tasks, the research utilizes tools from Nibabel [5], SimpleITK [6], Nilearn [7], Scipy [8], Scikit-Image [9], and TorchIO [10]. This paper is developed on top of Dr. Sreenivas's project for MRI segmentation task [11]. The dataset is taken from the Medical Segmentation Decathlon [12]. A V100 will be used to develop the project.

This research will primarily focus on augmentation and preprocessing. To better control the outcomes, we will manually perform all preprocessing and augmentation tasks with appropriate parameters. The resource-efficient U-Net model will be utilized, allowing the entire 3D image to fit in a GPU without the need to separate the volumetric image into slices or cubes. We have developed a special technique, Organ-Centric Resampling, which facilitates the preprocessing of the data. Additionally, Independent Random Fields Deformation will be applied to enhance image augmentation.

2 Related Works

The performance of 2D and 3D segmentation techniques is currently a subject of debate. Given the constraints of memory, there is a dilemma: should we sacrifice spatial connectivity to enhance performance on a per-slice basis, or should we prioritize spatial relationships, potentially at the cost of detection accuracy? The scoring method might be calculated on a per-slice basis for 2D input, which could differ from our project where the evaluation considers the entire segmentation in 3D.

We have come across research presenting a novel model called SWTR-Unet [13], a hybrid CNN architecture combining ResNet, transformer blocks, and a Unet-Style decoder, boasting a claimed dice score as high as 0.97. Furthermore, a sophisticated model named AIM-Unet [14] combines the Unet and Inception architectures. In this model, each output of the convolutional layer in the encoder part is combined with the output of the Inception module section, and this combined output is then transferred to the decoder. The paper claims a dice loss of 0.95. The above two papers report scores on the LiTS dataset. The following research utilizes an MRI-based dataset and employs the Unet++ [15] model, an enhanced version of the standard Unet model featuring a series of nested architectures and dense skip pathways, which should improve the flow of information and enhance gradient distribution throughout the network. However, this paper only achieved a dice score of 0.61. While the displayed slices may appear exceptionally clean, assessing their performance when stacked together remains challenging. What if adjacent slices fail to align seamlessly? How will this affect their appearance as a whole?".

While these papers demonstrate impressive performance, especially for tasks that rely on 2D representations, questions remain about their effectiveness when dealing with stacked slices to reconstruct an entire liver. Given that our objective is to work with the whole image, we will proceed using a 3D input.

As for models that support 3D input, we have found a paper introducing a novel technique known as Multiscale Cascaded Feature Attention (MCFA) [16], which utilizes the Attention-Gate Mechanism and claims to achieve a high Dice score of up to 0.955. Although using the same dataset, LiTS, this paper didn't mention whether the tumor is included in the output, and the image in the paper represents a single liver. It's possible that both the liver and tumor are merged. Since our output has 2 labels, tumor and liver, and we also use the argmax operation to process the output, it is not uncommon for the Dice score to be lower. This is because if the prediction of cancer, which is inside the liver, is inaccurate, the score for the liver will also be affected. This occurs because the operation assumes that one voxel can only have one label. The paper also conducted research on a regular 3D Unet, achieving a Dice score of 0.905. Another paper developed DALU-Net [17] and utilized a different liver dataset, boasting a score of 0.923. This project compares with 3D Unet, which achieved a score of 0.899. Comparing Dice scores with other papers is not the main objective of this project because different papers may utilize various methods to benchmark the Dice score. However, from the papers we observed, we will aim for a Dice score above 0.9.

3 Methodology

This section will elaborate on a special technique known as Organ-Centric Resampling, which will be enhanced by the application of Independent Random Fields Deformation, as well as the methods used for data preparation. Additionally, the two loss functions and metrics employed in this research will be explained. These combinations facilitate prolonged feature extraction, helping to mitigate early overfitting.

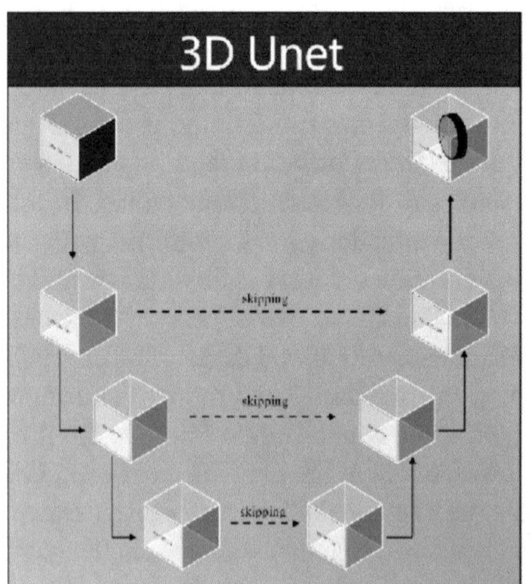

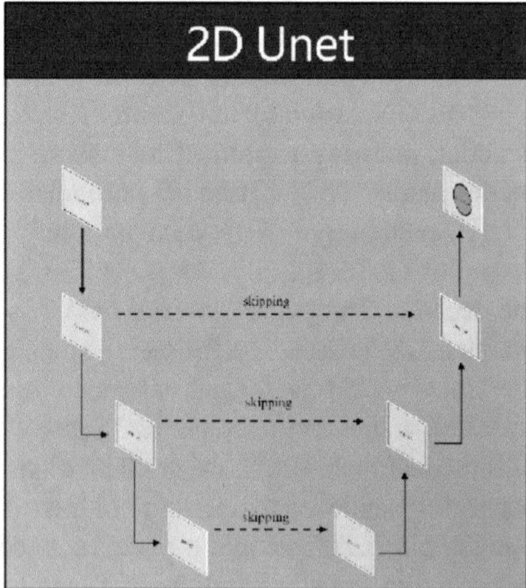

Fig. 1. The comparison between 3D Unet and 2D Unet architectures.

As observed in the workflows from Fig. 1., the slice entered into the 2D Unet workflow lacks connection with the whole image, leading the framework to overlook the correlation between slices in the 3D image. Another issue is that segmenting all 2D slices into an image and then joining them can result in a choppy outcome, as the slices lack a spatial relationship. Our approach processes the whole image without splitting it into patches. However, this comes at a cost: the input images are reduced to 128x128x128 to fit the training environment. The output will be a segmentation, consisting of a liver and tumors.

3.1 Preprocessing and Organ-Centric Resampling

It is not uncommon for the dataset to have some non-uniform files, especially with CT scans, as they could be gathered from various brands, each with their own formats and specifications. Our task is to ensure standardized input before delivering our images through the pipeline. These volumetric files are usually archived in either Dicom or Nibabel format. These files retain their headers, which record several pieces of information, on which researchers will focus, such as origin, spacing, and orientation. The origin informs us of the image's distance from the origin point, while spacing informs us about the anatomy of each voxel (a pixel in cube form). Slice thickness is the last element of spacing. Both origin and spacing can be found in the affine section of the header. The popular orientation is RAS (Right-Anterior-Superior), which defines the coordinate system for the image. Using the wrong orientation could lead to misinterpretation of the image, and it's essential for all images to have the same orientation, as the model will not be able to read the header. Orthogonality, the principle of ensuring that the unit is standardized as 1 unit, ensures that further spatial operations have the uniform and desirable outcome we expect. We will apply this standardization and then discard the header.

Cropping is very common, and researchers often use built-in tools which simply cut the image without consideration, potentially resulting in a cropped portion lacking liver labels. It's also possible that the cropped image may cut off a portion of the organ of interest. Such operations could affect the model's ability to localize the object we aim to segment. Our proposed method will take the label into account and then crop only that part, ensuring the liver is not cut. Additionally, the cropped image will position the organ in the middle, which is essential for further operations to operate without cutting the liver. The cropping process also diminishes compatibility issues with some loss functions, which will be explained further. Using hand-segmentation is feasible, but it might be challenging to aim for the middle part, not to mention the tediousness of the task.

A shown in Fig. 2., the "Preprocess and Organ-Centric Resampling" algorithm is meticulously designed to ensure high-quality, standardized medical imaging data. It corrects orientation and affine mismatches, normalizes image intensities within the range [–150, 250], and focuses on the organ of interest through automatic bounding box detection. This process intelligently expands the bounding box and crops it to a cube, ensuring the entire organ is captured without distortion or partial cuts, thus maintaining the liver's centrality in the images. Subsequently, images are resized to a uniform dimension, enhancing the dataset's consistency and reliability. By adjusting image properties to standard formats and rigorously checking orientation, spacing, and shape, the algorithm ensures data consistency, laying a robust foundation for precise and efficient downstream augmentation or analysis.

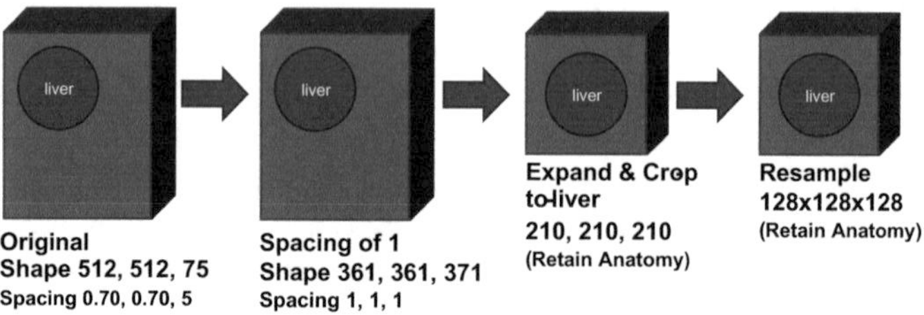

Fig. 2. Preprocess and organ-centric resampling workflow.

3.2 Loss Functions and Metrics

The loss function is a mathematical method used for measuring a model's performance. If the predictions match the ground truth, the gradients derived from the loss function will more significantly adjust the model's parameters that contributed to accurate predictions, while penalizing or leaving unchanged the parameters responsible for inaccuracies. We utilize a hybrid loss, which is not widely used as it can be challenging to adjust, and we must understand its nature to overcome its shortcomings. We might assign weights to it to prevent one component from over-emphasizing another, but this approach does not always work. We have tried two combinations: Categorical Focal Loss + Dice Loss, and Categorical Cross Entropy Loss + Dice Loss. There are many ways to implement these losses, but not all methods are optimized for volumetric data. We utilized the module from the project Segmentation Models 3D [1] to guarantee conformity. Dice loss is the opposite of the F-Score, and its objective is to minimize the loss during model training, using the following formula. Here, 'tp' is true positive, 'fp' is false positives, 'fn' is false negatives, and β determines the weight given to precision versus recall. If β is greater than 1, more weight will be given to recall, making the score more sensitive to false negatives. However, if β is less than 1, more weight will be given to precision, making the score more sensitive to false positives.

$$Loss_{Dice} = 1 - \frac{\left(1+\beta^2\right)\cdot tp + \beta^2 \cdot fn + fp}{\left(1+\beta^2\right)\cdot tp} \tag{1}$$

Categorical Focal Loss is an effective function for treating class imbalances, especially when the difference in labels is significant. 'Gt' is the ground truth label, 'pr' is the predicted probability, α is a weighting factor for classes to balance the importance of positive or negative examples, and log(pr) is the algorithm of the predicted probability. γ is a focusing parameter which adjusts the rate at which the importance of easy-to-classify samples is reduced. If γ is 0, the focal loss will become Categorical Cross Entropy Loss. As γ increases, the effect of loss on examples with high predicted probability is reduced, which helps the model to classify hard examples. The formula is as shown below:

$$Loss_{Catgorical\ Focal} = -gt \times \alpha \times \left(\left(1 - pr^{\gamma}\right) \times log(pr)\right) \tag{2}$$

Categorical Cross Entropy Loss is a general loss function that every researcher typically uses as the default, and it works fine in most circumstances. The 'gt' is the ground truth, and 'pr' is the predicted probability. The function focuses on how well each class is predicted, by penalizing predictions that deviate from the ground truth.

$$Loss_{Catgorical\ Cross\ Entropy} = -gt \times log(pr) \tag{3}$$

There are not many papers experimenting with these combinations of losses. Some have experimented with Exponential Logarithmic Dice (another form of Dice Loss) combined with weighted cross entropy loss and found that it's the best performing combination. Another study tried the combination of Binary Cross Entropy Loss and Dice Loss but found that its accuracy did not perform well on unseen data [18]. However, neither of them can be directly related to our work. We use categorical-based losses, which means that we are dealing with multiclass scenarios. Therefore, we decided to conduct our own experiments."

Our first combination was Categorical Focal Loss plus categorical cross entropy loss. We found that the loss didn't reduce as intended; it remained beyond the acceptable level of accuracy, varying but always exceeding 0.5. We suspect that this is due to the nature of our pre-processed data, which is cropped, causing it to become easier to classify with minimal class imbalances. This, at some level, might activate the penalty mechanism of the Focal loss itself. Therefore, we chose not to continue testing it.

Once we switched to Categorical Cross Entropy plus dice loss, we observed a tremendous change; the total loss went below 0.35. Therefore, we used this combination to test the effectiveness of augmentation afterwards.

We utilize only the IOU Metric, or Jaccard Index, for monitoring the training results, determining whether to continue, save, or terminate the training session. This metric does not affect the training optimization in any way. It quantifies the similarity and diversity of sample sets, which are suitable for the segmentation task. We set our training session to save only when the IOU metric shows an improvement, and this is an essential practice; otherwise, we might keep training and never know when to stop. Overfitting could occur without monitoring the training session. Our formula is as shown below. $|A \cap B|$ represents the common voxels between the predicted segmentation and the ground truth. $|A \cup B|$ is the count of all unique pixels found in either the predicted segmentation or the ground truth. It ranges from 0 to 1, where 1 means a perfect overlap.

$$IOU\ Metric = \frac{|A \cap B|}{|A \cup B|} \tag{4}$$

3.3 Independent Random Fields Deformation

In this process, we performed volumetric elastic deformation on an image and its corresponding mask. In the function, the alpha represents the intensity of deformation, while sigma is a component of the Gaussian filter, and it controls the smoothness of the deformation, with higher values resulting in smoother deformations. After trial and error, we determined that an adequate amount of alpha = 1000, sigma = 23 would not distort segmentation excessively. We learned that both alpha and sigma should be minimized to avoid excessive deformation of the label. Therefore, we modified how the formula performs an elastic deforms with Gaussian smoothened displacement field. We apply three randomized parameters and applied different values on each axis. The result is a more deformed image with minimal deformation parameters. Figure 3 shows an example of a slice of the displacement field to be applied to the image to deform both an image and its paired label.

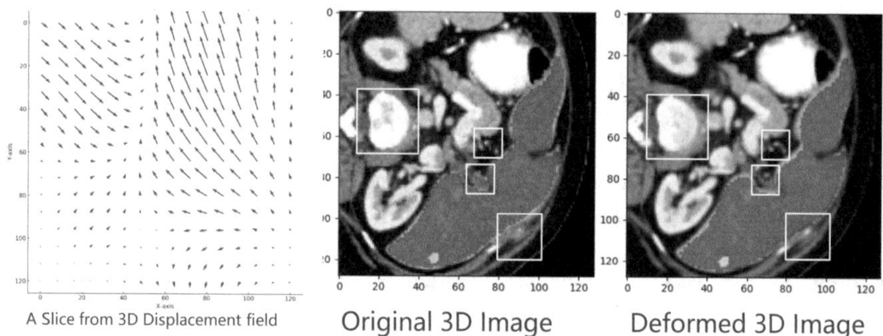

Fig. 3. A sample of a slice from a randomized 3D Displacement Field on the left, when applying to the middle original 3D image, the deformed 3D Image is created on the right.

4 Implementation and Experiments

Table 1. Dice scores for comparison with other setups. Note that other Dice scores were reported from the original previous work papers.

Dimensions	Model Name	Modality	Organ	Class	Dice
2D	SWTR*	CT	Liver	Liver	0.98
				Tumor	0.81
	AIM-Unet*	CT	Liver	Liver	0.95
				Tumor	0.75
	UNet++*	MRI	Liver	Liver	0.91
				Tumor	0.526
3D	MCFA*	CT	Liver	Tumor + liver	0.95
	DALU-Net*	CT	Liver	Tumor + liver	0.9
	Unet3D (ours)	CT	Liver	Tumor + liver	0.97
			Liver	liver	0.917
				Tumor	0.595

We implemented the callbacks method to automatically monitor the training, focusing on the increment of the IOU score; the model will be saved at every development.

We tested twice to check for the effectiveness of augmentation. When we disabled the augmentation, our model was trained for 59 epochs, achieving a Dice score of 0.76 for the liver and 0 for the cancer. However, once we enabled the augmentation, our model was trained for 151 epochs, and our Dice score increased to 0.917 for the liver and 0.595 for the cancer. Table 1 shows our results compared with results reported in other papers mentioned in the previous section. Figure 4 shows the output of our model with augmentations.

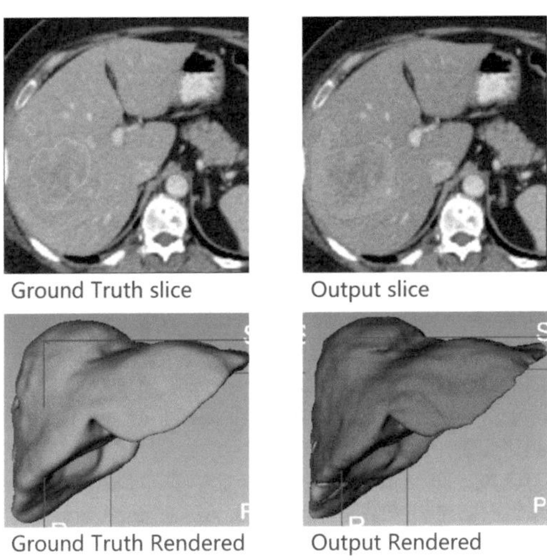

Fig. 4. The result of the model compared with the ground truth.

5 Summary

Our goal is to preserve spatial localization in segmentation, which necessitates the use of the full image as input. However, given that full input is memory exhaustive, we have faced the necessity to compromise with a simpler model. In this paper, we have demonstrated that a general model like U-Net can be significantly enhanced through strategic augmentations and preprocessing. In future research, we could apply this technique to other organs and explore more advanced modules to further optimize the model. This field is still in an early stage of study, and as computer hardware becomes increasingly advanced and more affordable, researchers are empowered to tackle the next level of challenges. Such advancements in medical imaging research have the potential to benefit everyone.

References

1. Set of models for segmentation of 3d volumes. https://github.com/ZFTurbo/segmentation_models_3D. Accessed 03 Oct 2023
2. Set of models for classifcation of 3d volumes. https://github.com/ZFTurbo/classification_models_3D. Accessed 03 Oct 2023
3. Keras (2015). https://keras.io. Accessed 03 Oct 2023
4. Tensorflow. https://www.tensorflow.org. Accessed 03 Oct 2023
5. Nibabel. http://nipy.org/nibabel. Accessed 05 Oct 2023
6. Simpleitk. https://simpleitk.org. Accessed 06 Oct 2023
7. Nilearn. https://nilearn.github.io. Accessed 06 Oct 2023
8. Scipy. https://www.scipy.org. Accessed 07 Oct 2023
9. Walt, V.D.S., et al.: The scikit-image contributors.: scikit-image: image processing in Python. PeerJ **2**, e453 (2014)
10. Pérez-Garcí, F., Sparks, R., Ourselin, S.: Torchio: a python library for efficient loading, preprocessing, augmentation and patch-based sampling of medical images in deep learning. Comput. Methods Programs Biomed. 106236 (2021)
11. Unet segmentation. https://github.com/bnsreenu/python_for_microscopists/tree/master/231_234_BraTa2020_Unet_segmentation. Accessed 03 Oct 2023
12. Simpson, A.L., et al.: A large annotated medical image dataset for the development and evaluation of segmentation algorithms (2019)
13. Hille, G., et al.: Joint liver and hepatic lesion segmentation in MRI using a hybrid CNN with transformer layers (2023)
14. Özcan, F., Ucan, O.N., Karacam, S., Tuncman, D.: Fully automatic liver and tumor segmentation from CT image using an aim-unet. Bioengineering **10**(2) (2023)
15. Wang, J., Peng, Y., Jing, S., Han, L., Li, T., Luo, J.: A deep-learning approach for segmentation of liver tumors in magnetic resonance imaging using unet++. BMC Cancer **23**(1) (2023)
16. Zhou, Y., Kong, Q., Zhu, Y., Su, Z.: Mcfa-unet: Multiscale cascaded feature attention u-net for liver segmentation. IRBM **44**(4), 100789 (2023)
17. Jeong, J.G., Choi, S., Kim, Y.J., Lee, W.S., Kim, K.: Deep 3D attention CLSTM u-net based automated liver segmentation and volumetry for the liver transplantation in abdominal CT volumes. Sci. Rep. **12**, 6370 (2022)
18. Galdran, A., Carneiro, G., Ballester, M.A.G.: On the optimal combination of cross-entropy and soft dice losses for lesion segmentation with out-of-distribution robustness. In: Yap, M.H., Kendrick, C., Cassidy, B. (eds.) Diabetic Foot Ulcers Grand Challenge. DFUC 2022. LNCS, vol. 13797, pp. 40–51. Springer, Cham (2023). https://doi.org/10.1007/978-3-031-26354-5_4

Monkeypox Lesion and Rash Stage Classification Using Deep Learning Technique

Orawan Chunhapran[1(✉)], Maleerat Maliyeam[1], and Gerald Quirchmayr[1,2]

[1] Faculty of Information Technology and Digital Innovation, King Mongkut's University of Technology North Bangkok, Bangkok, Thailand
chunhapran.o@email.kmutnb.ac.th, maleerat.m@itd.kmutnb.ac.th, Gerald.quirchmayr@univie.ac.th

[2] University of Vienna, Vienna, Austria

Abstract. Timely identification of Monkeypox symptoms stands as a key goal in averting a global pandemic. Furthermore, accurately assessing the disease stage and duration of Monkeypox rashes is imperative. The utilization of deep learning algorithms has become increasingly popular for the efficient diagnosis of diseases through medical imaging. In this research, the classification of Monkeypox lesions and rash stages is suggested through the application of deep learning techniques. The dataset comprises pictures of skin lesions and different stages of rashes. In the experiment, data augmentation methods were utilized to augment the sample size in both the training and testing sets. By employing a MobileNetV2 model, the outcome demonstrated an accuracy of 90.68% for pox and 90.62% for the pox rash stage. This approach aids healthcare professionals in crafting precise treatments tailored to the unique disease stage, offering accurate recommendations based on early-stage rash characteristics for each patient. Furthermore, it improves overall life expectancy by enabling more effective treatment strategies.

Keywords: Monkeypox · Rash Stage · Deep Learning Techniques · CNN

1 Introduction

The monkeypox virus has caused significant concerns in public health due to its rapid spread in more than 40 countries outside Africa, leading to a global alarm. Monkeypox is an infectious disease caused by the Zoonotic Orthopoxvirus, closely related to both monkeypox and smallpox and belongs to the Poxviridae family [1]. The monkeypox virus (MPXV) can spread through direct contact with the rash of an infected person or fluids from their body.

Additionally, humans can contract monkeypox from infected animals, either through bites or scratches or by preparing and consuming contaminated animal products [2, 3]. People infected with monkeypox can spread the disease from the onset of symptoms until the rash crusts over. The rash of monkeypox can be mistaken for other diseases like chickenpox, Herpes Zoster or syphilis. Diagnosis can be challenging in the early stages, relying on observing the initial abnormal features of the skin lesions and a history

P. Meesad et al. (Eds.): IC2IT 2024, LNNS 973, pp. 141–149, 2024.
https://doi.org/10.1007/978-3-031-58561-6_14

of exposure. Furthermore, confirmation can be made using polymerase chain reaction (PCR), a widely used method for diagnosing diseases, including COVID-19 [4–8]. When the skin lesions appear as red rashes, they indicate the rash phase, making early diagnosis a challenge for the general population to self-screen.

Diagnosing diseases using machine learning techniques, a subset of artificial intelligence, is becoming increasingly accepted in the medical field. Deep Convolutional Neural Networks (DCNNs) have achieved significant success in various medical applications. Additionally, studies have proposed deep learning techniques to classify monkeypox skin lesions. Md Manjurul Ahsan et al. [8] introduce a newly developed "Monkeypox2022" dataset that is publicly available to use and can be obtained from our shared GitHub repository. The dataset is created by collecting images from multiple open-source and online portals that do not impose any restrictions on use, even for commercial purposes, hence giving a safer path to use and disseminate such data when constructing and deploying any type of ML model, propose and evaluate a modified VGG16 model, which includes two distinct studies: Study One and Two. Our exploratory computational results indicate that our suggested model can identify Monkeypox patients with an accuracy of 97% and 86% for Study One and Two.

Although previous studies have successfully classified monkeypox using deep learning techniques, there have been studies that developed self-screening systems. For instance, Shams Nafisa Ali et al. [9] utilized the "Monkeypox Skin Lesion Dataset (MSLD)," comprising images of lesions from monkeypox, chickenpox, and measles. They employed the ResNet50 model and achieved an accuracy of 82.96%. It is not only the classification of monkeypox that is important. The duration of the rash is also interesting because it can predict the duration of symptoms.

According to our survey, there are studies that classify the occurrence of monkeypox [8, 9]. Monkeypox is a disease that is challenging to diagnose in its early stages because the symptoms in the human body resemble both measles and chickenpox. The rashes may sometimes be confused with those of chickenpox, starting from raised spots that turn into small fluid-filled bumps. These blister-like lesions eventually become scabs that may fall off later. The objective of this study is to clearly distinguish monkeypox lesions and rash stages from those of measles and chickenpox using deep learning techniques. The study utilizes two publicly available datasets, including a dataset of monkeypox lesion images and rash stages of monkeypox. The rash stages of monkeypox are classified into 5 stages: macules (1–2 days), papules (1–2 days), vesicles (3–5 days), pustules (5–7 days), and scabs (7–14 days).

2 Related Works

Monkeypox is a rare viral disease that closely resembles smallpox but is generally less severe. It is caused by the monkeypox virus, a member of the orthopoxvirus family. While the disease predominantly impacts monkeys, it can also infect humans, who can then transmit the virus to others through close contact. Several research studies have proposed the use of deep learning technology to identify skin lesions from monkeypox, employing image classification as a notable methodology.

Muhammed Coşkun Irmak et al. [10] classification performed using the pre-trained CNN networks MobileNetV2, VGG16, and VGG19 on the Monkeypox Skin Image

Dataset. The highest performance scores were obtained with MobileNetV2, with 91.38% accuracy.

Selen Gürbüz and Galip Aydın [11] conducted a study on image analysis for detection is highly significant. They employed five pre-trained network architectures and transfer learning approaches, including DesNet121, ResNet50, Xception, EfficientNetB3, and EfficientNetB7, to classify skin lesions caused by the monkeypox virus. The study's results revealed that EfficientNetB7 exhibited the highest performance score, achieving a detection accuracy rate of 90%. This indicates superior accuracy in detecting skin lesions caused by the monkeypox virus.

Murat Altun et al. [12] conducted a study on monkeypox in a potential case of spread through a skin rash, using a fast and secure deep learning method. They applied this method to customized models, including MobileNetV3-s, EfficientNetV2, ResNET50, Vgg19, DenseNet121, and Xception. The MobileNetV3 model, customized for optimal performance, achieved an accuracy of 0.96.

Md Manjurul Ahsan et al. [8] introduced the "Monkeypox2022" dataset, which is now publicly available and can be accessed from their shared GitHub repository. This dataset was curated by gathering images from various open sources and online portals without any usage restrictions. The study evaluated the VGG16 model, demonstrating its ability to identify monkeypox patients with an accuracy of 97%.

Kapil Sharma et al. [13] proposed the ResNet model and then compared the results of VGG 16, Inception V3 and CapsuleNet to classify various Measles, ChickenPox and MonkeyPox. The highest performance scores obtained with a custom RESNET-18-based model achieved an accuracy of 84.59% in the classification of monkeypox, measles and chickenpox.

Md. Enamul Haque et al. [14] implemented an architecture comprising ception-CBAM-Dense layers, which outperformed other models in classifying monkeypox and various diseases, achieving a validation accuracy of 83.89%. The study included a comparative analysis involving five deep learning models: VGG19, Xception, DenseNet121, EfficientNetB3, and MobileNetV2, all integrated with channel and spatial attention mechanisms.

From the literature review, several research studies [10, 12, 14] have utilized MobileNet-v2, yielding efficient results. This has inspired researchers to incorporate the MobileNet-v2 architecture into their own studies. MobileNet-v2 is a deep neural network with 53 layers that has been trained on over a million images from the ImageNet database [15]. The pre-trained network can accurately classify images into 1,000 object categories. One of the key advantages of MobileNet-v2 is its resource efficiency, making it well-suited for deployment on devices with limited capacity, such as mobile phones or IoT devices with constraints in processing power and storage space. Furthermore, the accuracy in object categorization remains high, making it beneficial for applications that require precision, such as image classification and object analysis in various fields. These advantages suggest potential adaptations in the future.

These investigations highlight the potential of utilizing classification methods, particularly deep learning, to effectively identify monkeypox in images depicting skin lesions. Such techniques offer valuable tools for the rapid and accurate detection of the disease, playing a crucial role in preventing its transmission and safeguarding public health.

3 Research Methodology

The section presents a systematic research methodology, which includes the proposed framework, dataset, data augmentation techniques, classification, and performance measurement.

3.1 Proposed Framework

The proposed framework of Monkeypox Lesion and Rash Stage Classification using Deep Learning, shown in Fig. 1, there are two parts of classification. The first part is used to classify Monkeypox Lesions and the second part is used to classify the level of rash stage. Both parts use the same image and the result of each part will be combined to give the final output.

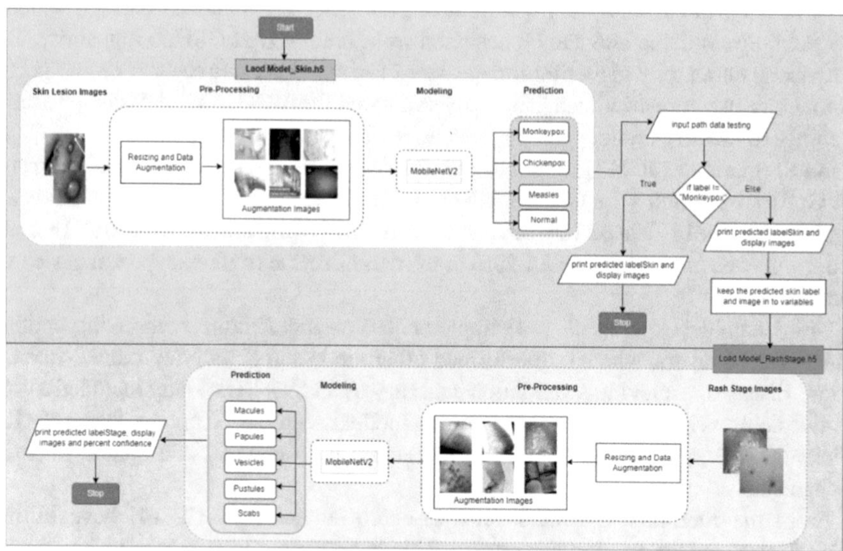

Fig. 1. Proposed framework.

3.2 Dataset

Two datasets of skin images used in the experiment: (1) the Skin Lesion Image dataset and (2) the Rash Stages Image dataset.

1) **Skin Lesion Image Dataset**
   This dataset was developed by the Department of Computer Science and Engineering at Islamic University, Kushtia-7003, Bangladesh [16]. This dataset consists of 4 classes: Monkeypox class contains 279 images, Chickenpox contains 107 images, Measles contains 91 images and Normal contains 293 images. Figure 2 shows some example images from our experiment dataset.

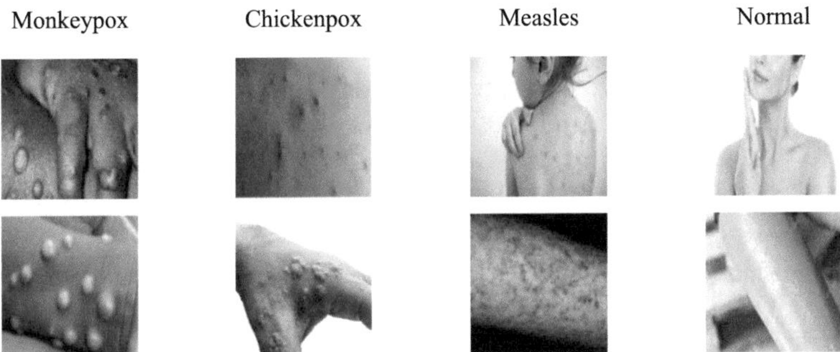

Fig. 2. Sample skin images of example skin images of Monkeypox, Chickenpox, and healthy cases from the experiment dataset [16]

2) **Rash Stages Image Dataset**
The dataset includes images of lesions or pustules that appear on the skin during different stages of monkeypox. These images were made publicly available by various health organizations, including the Centers for Disease Control and Prevention (CDC) and the National Health Service (NHS) in the United Kingdom. The dataset is designed to differentiate between different stages of monkeypox and estimate the duration of the disease [17]. The dataset consists of five stages: Macules (1–2 days) contain 71 images, Papules (1–2 days) contain 52 images, Vesicles (3–5 days) contain 48 images, Pustules (5–7 days) contain 120 images and Scabs (7–14 days) contains 35 images. Figure 3 shows an example rash stage or lesion of monkeypox and the duration of the rash.

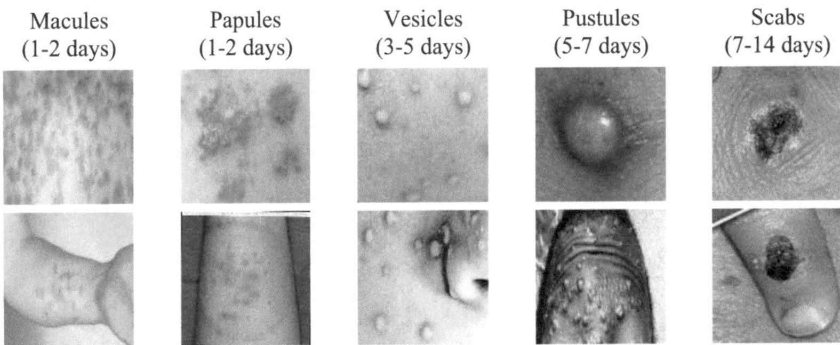

Fig. 3. Sample rash stages of monkeypox and the duration of the rash [17].

3.3 Data Augmentation Techniques

This study primarily aims to differentiate between monkeypox cases and those that may be confused with it, such as chickenpox. Therefore, various data augmentation methods, including blur, noise, brightness, darkness, and cropping, have been employed to enhance the classification performance.

3.4 Classification

The dataset was split into 80% training set and 20% testing data. Augmentation techniques were applied to the training data to enhance dataset diversity and improve the predictive capability of the model. The augmented data, combined with the original training set, was fed into the pre-trained MobileNetV2 model in TensorFlow, capable of recognizing 1000 classes of images. Transfer learning was utilized to reduce the required training data and expedite the training process. This methodology was also employed to retrain the model for recognizing classes in a custom dataset. For instance, the retrained model was effective in distinguishing between various skin lesions and rash stages.

3.5 Performance Measure

The model evaluation such as accuracy, sensitivity, and specificity were used to evaluate the performance of the CNN model. The calculation types of the metrics are shown in Eqs. (1)–(3), where TP, FN, FP, and TN represent the number of true positives, false negatives, false positives and true negatives.

$$\text{Sensitivity} = \frac{TP}{TP + FN} \tag{1}$$

$$\text{Specificity} = \frac{TN}{TN + FP} \tag{2}$$

$$\text{Accuracy} = \frac{TP + TN}{TP + FP + TN + FN} \tag{3}$$

4 Experimental and Results

This study focuses on distinguishing cases of Monkeypox from cases that could be mistaken for it. Therefore, there is an analysis and classification of the stages of the monkeypox rash using deep learning techniques.

4.1 Experimental Design

1. **Computational Configuration**
   Using the Keras library, the network architectures were de-developed for the purpose of training the skin lesion categorization. The Kaggle platform offered GPU assistance to accelerate the architectures, which were updated throughout training. The adaptive learning rate optimizer with a BinaryCrossentropy loss function was used for training, and the Adam optimizer with an initial learning rate of 0.001 was used as the starting hyperparameter setting. A typical batch size of 32 offered a fair compromise between model convergence and computing efficiency. A total of 100 epochs were used for the training.

   Additionally, the optimized structure of the MobileNetV2 transfer learning model is meticulously designed to ensure exceptional performance. Leveraging the Rectified

Linear Unit (ReLU) activation function, we have configured the ideal number of intermediate layers to be 256, 128, and 64.

For the rash stage, once the skin lesion classification predicts Monkeypox, additional predictions are carried out to discern the specific rash stages associated with Monkeypox. These stages encompass Macules (1–2 days), Papules (1–2 days), Vesicles (3–5 days), Pustules (5–7 days), and Scabs. The configuration settings for these predictions align with those employed in the initial skin lesion classification process.

2. **Model Deployment**
In the process of predicting the skin disease type as "monkeypox," relevant variables linked to the skin disease image and group name are stored using the classification of rash stages since the disease's onset. These stages include macula (1–2 days), papule (1–2 days), cysts (3–5 days), pustules (5–7 days), and scabs. This approach is implemented to support an initial diagnosis of monkeypox and to categorize the various stages of the monkeypox rash.

4.2 Results

The result of this study were generated using the Python programming language in Google Colaboratory. A dataset comprising 1,766 images of skin lesions was utilized, with a split of 1,270 images for training, 354 images for testing, and 142 images for validation in the classification of "Monkeypox." The test results exhibited a precision of 91%, recall of 91%, f1-score of 91%, and an accuracy of 90.68%, Sensitivity of 91%, and specificity of 91%. Figure 4 shows the confusion matrix of skin lesion classification.

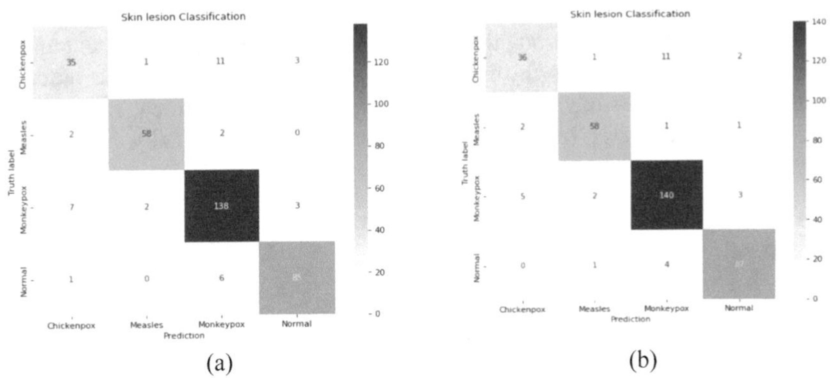

Fig. 4. The comparison between the initial model and the optimized structure of MobileNetV2 for skin lesion classification (a) represents the pre-trained MobileNetV2, (b) showcases the optimized structure of MobileNetV2.

The testing accuracy for pre-training MobileNetV2 was 89.27%, while the optimized structure of MobileNetV2 achieved a testing accuracy in the range of 90.68%.

In the testing phase for rash status, a total of 959 images were examined for the classification of rash stages. The test outcomes demonstrated an accuracy of 91%, recall

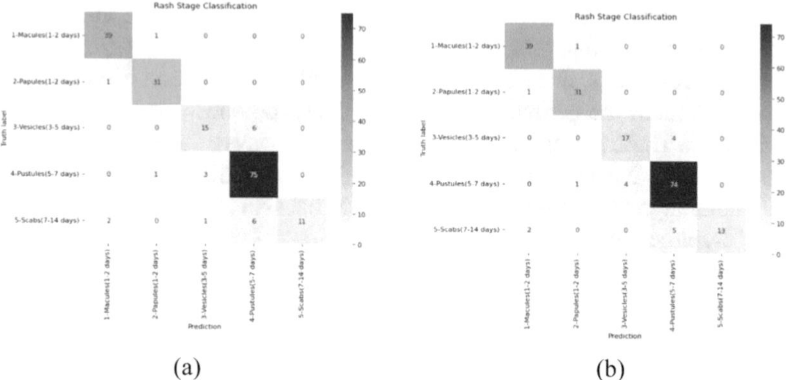

(a) (b)

Fig. 5. The comparison between the initial model and the optimized structure of MobileNetV2 for rash stage classification (a) represents the pre-trained MobileNetV2, (b) showcases the optimized structure of MobileNetV2.

of 91%, f1 score of 91%, and an accuracy of 90.62%, Sensitivity of 91%, and specificity of 90%. Figure 5 shows the confusion matrix of rash stage lesion classification.

The testing accuracy for pre-training MobileNetV2 was 89.06%, while the optimized structure of MobileNetV2 achieved a testing accuracy in the range of 90.62%.

5 Conclusion

Utilizing deep learning methods for categorizing monkeypox lesions and rash stages has demonstrated efficacy in promptly identifying patients and managing community transmission, attaining a noteworthy operational accuracy of 90%. Despite the numerous benefits offered by this study, certain limitations have been identified. Primarily, the training dataset comprises a restricted number of samples. Additionally, there is room for enhancing the system's accuracy.

References

1. McCollum, A.M., Damon, I.K.: Human Monkeypox. Clin. Infect. Dis. **58**(2), 260–267 (2013)
2. Farahat, R.A., et al.: Monkeypox and human transmission: are we on the verge of another pandemic? Travel Med. Infect. Dis. **49**, 102387 (2022)
3. Adler, H., et al.: Clinical features and management of human Monkeypox: a retrospective observational study in the UK. Lancet Infect. Dis. **22**(8), 1153–1162 (2022)
4. Ahsan, M.M., Gupta, K.D., Islam, M.M., Sen, S., Rahman, M.L., Shakhawat Hossain, M.: COVID-19 symptoms detection based on NASNETMOBILE with explainable AI using various imaging modalities. Mach. Learn. Knowl. Extr. **2**(4), 490–504 (2020)
5. Ahsan, M.M., Alam, E.T., Trafalis, T., Huebner, P.: Deep MLP-CNN model using mixed-data to distinguish between COVID-19 and non-COVID-19 patients. Symmetry **12**(9), 1526 (2020)
6. Ahsan, M.M., et al.: Detecting SARS-CoV-2 from chest X-Ray using artificial intelligence. IEEE Access **9**, 35501–35513 (2021)

7. Ahsan, M.M., Nazim, R., Siddique, Z., Huebner, P.: Detection of COVID-19 patients from CT scan and chest X-ray data using modified mobilenetv2 and lime. Healthcare **9**(9), 1099 (2021)
8. Ahsan, M.M., Uddin, M.R., Farjana, M., Sakib, A.N., Momin, K.A., Luna, S.A.: Image Data collection and implementation of deep learning-based model in detecting Monkeypox disease using modified VGG16. arXiv, abc/2206.01862. (2022)
9. Ali, S.N., et al.: Monkeypox skin lesion detection using deep learning models: a feasibility study (2022). arXiv:2207.03342
10. Irmak, M.C., Aydin, T., Yağanoğlu, M.: Monkeypox Skin Lesion Detection with MobileNetV2 and VGGNet Models. Medical Technologies Congress (TIPTEKNO), Antalya, Turkey, 1–4 (2022)
11. Gürbüz, S., Aydin, G.: Monkeypox skin lesion detection using deep learning models. In: International Conference on Computers and Artificial Intelligence Technologies (CAIT), pp. 66–70. Quzhou, China (2022)
12. Altun, M., Gürüler, H., Özkaraca, O., Khan, F., Khan, J., Lee, Y.: Monkeypox detection using CNN with transfer learning. Sensors **23**(4), 1783 (2023)
13. Sharma, K., Kumar, K.V., Mittal, M.: MonkeyPox, measles and chickenpox detection through image-processing using residual neural network (ResNet). In: 6th International Conference on Information Systems and Computer Networks (ISCON), Mathura, India, 2023, pp. 1–6 (2023)
14. Haque, M.E., Ahmed, M.R., Nila, R.S., Islam, S.: Classification of Human Monkeypox Disease Using Deep Learning Models and Attention Mechanisms (2022). arXiv:2211.15459
15. ImageNet. http://www.image-net.org. Accessed 28 Sep 2022
16. Bala, D.: Monkeypox Skin Images Dataset (MSID), Kaggle. https://www.kaggle.com/datasets/dipuiucse/monkeypoxskinimagedataset. Accessed 11 Jan 2022
17. Centers for Disease Control and Prevention. https://www.cdc.gov/poxvirus/monkeypox/clinicians/clinical-recognition.html. Accessed 28 Sep 2022

Durian Ripeness Classification Using Deep Transfer Learning

Santi Sukkasem(✉), Watchareewan Jitsakul, and Phayung Meesad

Faculty of Information Technology and Digital Innovation, King Mongkut's University of Technology North Bangkok, Bangkok, Thailand
s6407021910024@email.kmutnb.ac.th, {watchareewan.j, phayung.m}@itd.kmutnb.ac.th

Abstract. This study aims to investigate the application of deep transfer learning to the ripe-ness levels of the Monthong durian classification. The objective is to compare the effectiveness of different classification models, including Inception-v2, Inception-v3, VGGNet16, VGGNet19, ResNet-50, ResNet-101, ResNet-152, MobileNet-v2, and MobileNet-v3. The dataset for this research consists of 1000 images of the Monthong durian, divided into 4 levels: over-ripe, semi-ripe, unripe, and ripe. Hyperparameters were specified to enhance the accuracy of classification of the ripeness levels using pretrained and modified models. The deep transfer learning models and the top 3 performing models were MobileNet-v2 with an accuracy of 95.50%, VGG16 with an accuracy of 94.50%, and VGG19 with an accuracy of 94.50%. The deep transfer learning model with the lowest performance was Inception-v2, achieving an accuracy of 81.50%, particularly in identifying the complex characteristics of the Monthong durian. We anticipate that the results of this study will not only contribute to advancements in durian ripe-ness classification but also provide valuable insights for stakeholders in the durian industry. This will guide cultivation practices and marketing strategies with more comprehensive data while examining how this research contributes to the classification of agricultural practices and the application of machine learning in a broader range of industrial sectors.

Keywords: Image Processing · Convolutional Neural Network (CNN) · Deep Transfer Learning · Durian

1 Introduction

Durian, also known as the "King of Fruits," holds a prominent status in the world of gastronomy, captivating taste buds with its unique flavor and aroma. Beyond its culinary appeal, the durian industry is increasingly recognizing the significance of understanding and classifying the levels of satisfaction associated with this tropical fruit. The concept of durian happiness classification involves distinguishing the quality and satisfaction experienced by consumers and distributors [1], considering factors such as taste, texture, peel, and thorns [2]. Efficient categorization in the durian industry is crucial for

P. Meesad et al. (Eds.): IC2IT 2024, LNNS 973, pp. 150–161, 2024.
https://doi.org/10.1007/978-3-031-58561-6_15

several reasons. First, it helps producers and distributors respond efficiently to consumer demand, facilitating supply chain management and ensuring customer satisfaction [3]. Second, identifying ripeness levels contributes to the continual improvement of durian cultivation practices. To enhance the overall quality of the fruit, an effective classification is most efficient [4]. This research focuses on deep transfer learning, an advanced machine learning technique that leverages knowledge gained in one task to improve performance in another [5]. Deep transfer learning has demonstrated remarkable success in various domains, yet its potential to classification the levels of ripeness associated with durian remains largely unexplored. The study aims to explore and evaluate the performance of a deep transfer learning model for accurately classification ripeness levels in durian.

The main objective is to compare the performance of different classification models using deep transfer learning techniques, including GoogleNet, and MobileNet. This comparative analysis seeks to provide insights that not only improve the accuracy of classification of durian ripeness levels but also contribute valuable knowledge to broad agricultural classification using advanced machine learning methods. The aim is to lay a foundation for optimizing practices within the durian industry and promoting innovation in the assessment of optimal fruit quality.

2 Literature Reviews

2.1 Convolutional Neural Networks Architecture (CNN)

Convolutional Neural Networks (CNNs) are a specific type of deep artificial neural network designed specifically for computer vision tasks. The process of image recognition and feature extraction in CNNs involves several key components that enable efficient processing and learning of hierarchical representations from image data. Yann LeCun and his team [6] developed the first CNN in 1994. In 1998, Y. LeCun and his team [7] proposed LeNet as the foundational architecture for CNNs in the ImageNet project. Some CNN architectures, such as GoogleNet, VGGNet, ResNet, and MobileNet, have more layers than LeNet5 and collectively possess over 60 million parameters. In comparison, LeNet only has 60,000 parameters for "transfer learning."

The principle of inception modules involves using multiple convolutions of different sizes to process the same input and then concatenating them together. This methodenables the model to perform multi-level feature extraction at each step, allowing it to learn features at different scales. For instance, a 5×5 convolution would capture the overall features of the data, whereas a 3×3 convolution would focus on smaller, more detailed features [8].

The architecture of VGGNet is predominantly composed of 3×3 convolutions. Using smaller filters increases non-linearity in the network, thereby increasing the degrees of freedom, which justifies this choice. Employing multiple layers of small filters allows the model to capture finer details. For example, using two layers of 3×3 filters enables the model to perceive a 5×5 area, and using three layers allows it to perceive a 7×7 area. This leads to a greater amount of information while reducing the num the user's text is [9].

The ResNet architecture incorporates identity maps, which are element wise additions, every two layers. This technique has been proven to aid in gradient propagation, as it allows errors from backpropagation to be effectively transmitted over multiple paths. Alternatively, from another perspective, this technique helps combine features at multiple levels, similar to the inception modules [10].

The MobileNet is a model specifically designed for mobile devices, accompanied by pre-trained weights derived from the ImageNet database, which contains millions of images. This model is versatile and capable of performing many tasks. MobileNet [11] sacrifices some performance compared to other models in order to minimize resource usage, making it a resource-efficient design.

2.2 Durian Ripeness Classification

Classifying durian ripeness levels poses challenges. Traditional sensory methods by experts can lack consistency. Machine learning models may require customization based on subjective preferences and scarce labeling of durian-specific ripeness levels [12]. The variability in subjective perception of flavor and texture complexity necessitates complex models [13]. Deep transfer learning addresses these limitations by enabling knowledge transfer from related domains, improving generalization and adaption to durian classification intricacies [14].

Beyond precision measurement, performance factors like computation speed, model durability, and dataset adaptability are critical. Deep transfer learning has increased performance in other domains, making it an attractive alternative to address durian ripeness classification challenges. By leveraging knowledge from broader domains and narrowing focus to address durian complexity [15], deep transfer learning can create efficient and accurate models.

As ripeness level classification relates to durian consumption and exportation, deep transfer learning's versatility makes it a promising approach for this multifaceted task. Its knowledge transfer capacities allow customization based on individual preferences while also improving consistency over sensory-dependent methods. By bridging broad and specific domains, deep transfer learning can aptly manage the nuances of classifying durian ripeness.

2.3 Related Work

Deep transfer learning plays a crucial role in classifying the ripeness levels of Monthong durians, highlighting the significant role of convolutional neural networks (CNN) in such analyses. The utility of deep learning extends to various agricultural applications. For example, Hongboonmee and Jantawong [16] explored the application of deep learning techniques to assess the sweetness level of watermelons using smartphone images. They compared the efficacy of two image classification models, InceptionV3 and MobileNet, across four watermelon image categories based on variety and sweetness level, with a dataset comprising 400 images. Both models demonstrated comparable performance, achieving an impressive accuracy of 97.20%.

Chaikaew et al. [17] proposed an innovative use of artificial neural networks for automating pineapple sorting by skin color, thus classifying pineapples into unripe, partially ripe, and fully ripe categories. This approach significantly minimized the need for manual labor, achieving prediction accuracies of 79.00% for unripe, 82.00% for partially ripe, and 100% for fully ripe pineapples.

Gallenero and Villaverde [18] addressed the pressing issue of durian leaf blight, a significant agricultural challenge in Southeast Asia, particularly in the Philippines. Utilizing the MobileNet CNN model within the Duri Premium application, Gallenero achieved a diagnostic accuracy of 93.33% for identifying various conditions affecting durian leaves, demonstrating the potential of mobile-based CNN applications in plant disease management.

In the field of remote sensing, Alem and Kumar [19] showcased how deep learning could enhance the analysis and interpretation of remote sensing images for Land Cover Land Use (LCLU) classification. Their methodology involved either developing new deep learning models or utilizing pre-trained networks, highlighting the adaptability and efficacy of deep learning in environmental monitoring.

Amin et al. [20] demonstrated the superiority of the Xception-based model in detecting stress levels among drivers, outperforming other models such as GoogLeNet and DenseNet-201. This study illustrates the advancements in deep learning for behavioral and psychological analysis.

Aranha et al. [21] employed convolutional neural networks, refined through transfer learning, on a dataset of 38,727 high-quality retinal images. Their work achieved significant accuracy in classifying conditions such as cataract and diabetic retinopathy, showcasing the potential of deep learning in medical diagnostics.

Miraei Ashtiani et al. [22] tackled the challenging task of classifying mulberry fruit ripeness using CNNs, achieving up to 98.65% accuracy with models like AlexNet and ResNet-18. This study underscores the advantages of transfer learning in enhancing classification accuracy.

In our research, we aim to further the field of Monthong durian ripeness classification through deep transfer learning algorithms. We examine a variety of models, including GoogleNet (Inception-v2 and Inception-v3), VGGNet (VGGNet16 and VGGNet19), ResNet (ResNet-50, ResNet-101, and ResNet-152), and MobileNet (MobileNet-v2 and MobileNet-v3), assessing their performance in this specific context. This effort builds upon the foundation laid by previous studies [16–22], contributing to a deeper understanding of the potential of deep transfer learning in accurately classifying the ripeness of agricultural products and rein-forcing the versatility of CNNs across a range of applications.

3 Research Methodology

Process for classification durian ripeness levels By using deep transfer learning. To train the dataset with the trained model below. Then the model is tested with test images. This study was conducted using a model consisting of GoogleNet, including Inception-v2, Inception-v3, VGGNet, including VGGNet16, VGGNet19, ResNet parts, including ResNet-50, ResNet-101, and ResNet-152, and MobileNet parts, including MobileNet-v2, MobileNet-v3. To classify the level of ripeness of durian, which is divided into 4 levels, including unripe, semi-ripe, ripe, and over-ripe, here is a JPG photo of Monthong durian fruit with a resolution of 224 $\times$ 224 pixels (Fig. 1).

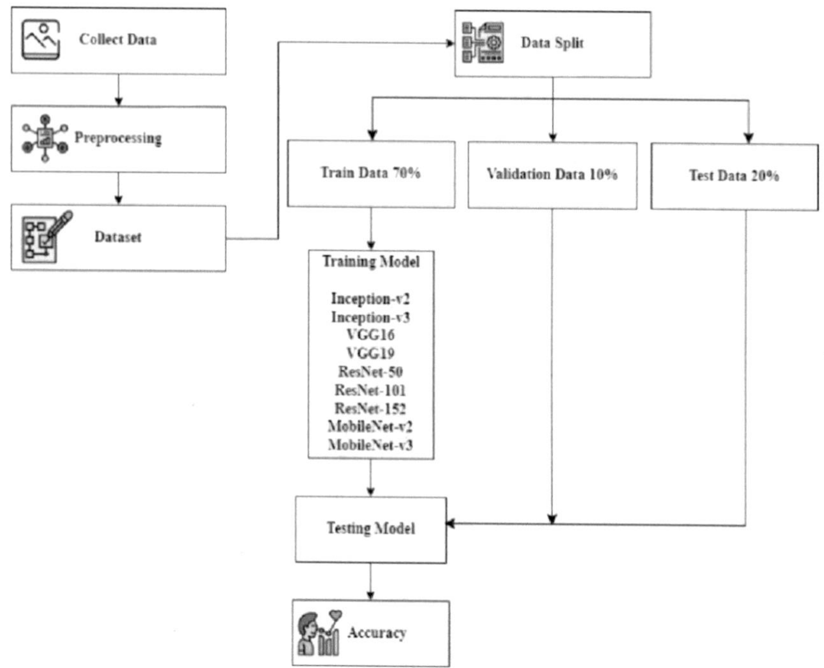

Fig. 1. Conceptual framework for research.

3.1 Collect Data

The data used in this experiment was photographed in the Monthong durian cultivation area of Tambon Tham Talu, Ban Nang Sata District, Yala Province. The investigation focused on determining the ripeness stages of the Monthong durian from the initial growth stage to the harvesting stage. Each set of data consists of training data, validation data, and testing data, each with corresponding images. The dataset consists of 1,000 photographs, with every class having the same 250 photographs, including 70% training data, 10% validation data, and 20% testing data. The dataset is divided into 4 levels of ripe-ness classification: overripe = class 0, semi-ripe = class 1, unripe = class2, and ripe = class 3 (Fig. 2).

Fig. 2. Levels of ripeness of monthong durian.

Use deep transfer learning techniques with the following methods: GoogleNet, including Inception-v2, Inception-v3, VGGNet, including VGGNet16, VGGNet19, ResNet parts, including ResNet-50, ResNet-101, ResNet-152, and MobileNet parts, including MobileNet-v2 and MobileNet-v3 [23].

3.2 Details of Image Preparation for Deep Transfer Learning

A robust dataset of high-quality, diverse durian images is crucial for successful deep transfer learning. Images are standardized to 224×224 pixels and augmented through transformations like rotation and flipping to generalize the model and prevent overfitting.

The data is systematically divided into training, validation, and testing sets. The training set educates the model, while the validation set tunes hyperparameters and evaluates interim performance. The testing set ultimately assesses accuracy on unseen data.

Pixel value normalization further standardizes data for efficient learning. Pre-trained models like GoogleNet and ResNet allow transfer learning by feature extraction or fine-tuning to leverage existing image recognition capabilities.

Careful training monitoring avoids overfitting while refining parameters for optimization. Rigorous validation tuning precedes final testing to evaluate predictive accuracy on new data. This end-to-end approach enables an accurate and reliable model for durian ripeness classification through deep transfer learning.

Data quality, diversity and standardization provide a robust starting foundation. Strategic dataset splitting reduces overspecialization. Transfer learning and iterative refinement boosts model versatility for durian-specific performance. Holistic validation ensures generalizability before final capability assessment.

3.3 Hyperparameters Setup

(See Fig. 3).

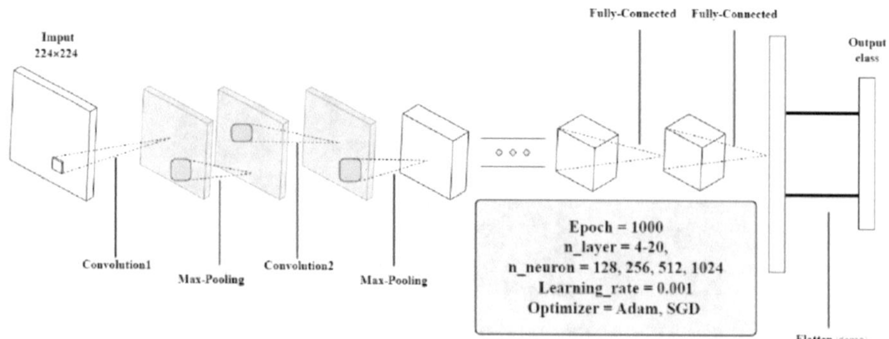

Fig. 3. CNN architecture and hyperparameter.

The hyperparameters of the CNN model all models set the same hyperparameter values, including epoch = 1000, n_layer = 4–20, n_neuron = 128, 256, 512, 1024, learning rate = 0.001, and performance-enhancing methods = Adam, SGD.

4 Experimental Results

Our work comparison deep transfer learning to classify Durian Ripeness. The result is shown in Tables 1, 2, 3, 4, 5, 6, 7, 8 and 9.

Table 1. Report Inception-v2.

Parameter		Class	Precision	Recall	F1-score
Epoch	1000	Overripe (0)	60.87	89.36	72.41
n_layer	18	Semi-ripe (1)	95.45	89.36	92.31
n_neuron	1024	Unripe (2)	93.55	52.73	67.44
learning_rate	0.001	Ripe (3)	89.29	98.04	93.46
optimizer	Adam	Accuracy = 81.50			

Table 1 shows the best result of Inception-v2 with Epoch = 1000, n_layer = 18, n_neuron = 1024, learning_rate = 0.001 and optimizer = Adam. The result was accuracy of 81.50%, precision of 84.79%, recall of 82.37%, and f1-score of 81.41%.

Table 2 shows the best result of Inception-v3 with Epoch = 1000, n_layer = 15, n_neuron = 512, learning_rate = 0.001 and optimizer = Adam. The result was accuracy of 90.00%, precision of 90.58%, recall of 89.95%, and f1-score of 90.11%.

Table 2. Report Inception-v3.

Parameter		Class	Precision	Recall	F1-score
Epoch	1000	Overripe (0)	79.25	89.36	84.00
n_layer	15	Semi-ripe (1)	97.62	87.23	92.13
n_neuron	512	Unripe (2)	87.50	89.09	88.29
learning_rate	0.001	Ripe (3)	97.96	94.12	96.00
optimizer	Adam	Accuracy = 90.00			

Table 3. Report VGG16.

Parameter		Class	Precision	Recall	F1-score
Epoch	1000	Overripe (0)	84.91	95.74	90.00
n_layer	17	Semi-ripe (1)	95.92	100	97.92
n_neuron	512	Unripe (2)	97.87	83.64	90.20
learning_rate	0.001	Ripe (3)	100	100	100
optimizer	Adam	Accuracy = 94.50			

Table 3 shows the best result of VGG16 with Epoch = 1000, n_layer = 17, n_neuron = 512, learning_rate = 0.001 and optimizer = Adam. The result was accuracy of 94.50%, precision of 94.67%, recall of 94.85%, and f1-score of 94.53%.

Table 4. Report VGG19.

Parameter		Class	Precision	Recall	F1-score
Epoch	1000	Overripe (0)	93.02	85.11	88.89
n_layer	17	Semi-ripe (1)	100	100	100
n_neuron	512	Unripe (2)	86.67	94.55	90.43
learning_rate	0.001	Ripe (3)	100	98.04	99.01
optimizer	Adam	Accuracy = 94.50			

Table 4 shows the best result of VGG19 with Epoch = 1000, n_layer = 17, n_neuron = 512, learning_rate = 0.001 and optimizer = Adam. The result was accuracy of 94.50%, precision of 94.92%, recall of 94.42%, and f1-score of 94.51%.

Table 5 shows the best result of ResNet-50 with Epoch = 1000, n_layer = 8, n_neuron = 512, learning_rate = 0.001 and optimizer = Adam. The result was accuracy of 87.00%, precision of 87.09%, recall of 87.22%, and f1-score of 87.07%.

Table 5. Report ResNet-50.

Parameter		Class	Precision	Recall	F1-score
Epoch	1000	Overripe (0)	84.09	78.72	81.32
n_layer	8	Semi-ripe (1)	95.74	95.74	95.74
n_neuron	512	Unripe (2)	79.25	76.36	77.78
learning_rate	0.001	Ripe (3)	89.29	98.04	93.46
optimizer	Adam	Accuracy = 87.00			

Table 6. Report ResNet-101.

Parameter		Class	Precision	Recall	F1-score
Epoch	1000	Overripe (0)	76.47	82.98	79.59
n_layer	14	Semi-ripe (1)	91.84	95.74	93.75
n_neuron	512	Unripe (2)	81.82	65.45	72.73
learning_rate	0.001	Ripe (3)	87.50	96.08	91.59
optimizer	Adam	Accuracy = 84.50			

Table 6 shows the best result of ResNet-101 with Epoch = 1000, n_layer = 14, n_neuron = 512, learning_rate = 0.001 and optimizer = Adam. The result was accuracy of 84.50%, precision of 84.41%, recall of 85.06%, and f1-score of 84.41%.

Table 7. Report ResNet-152.

Parameter		Class	Precision	Recall	F1-score
Epoch	1000	Overripe (0)	76.36	89.36	82.35
n_layer	8	Semi-ripe (1)	95.56	91.49	93.48
n_neuron	512	Unripe (2)	86.36	69.09	76.77
learning_rate	0.001	Ripe (3)	91.07	100	95.33
optimizer	Adam	Accuracy = 87.00			

Table 7 shows the best result of ResNet-152 with Epoch = 1000, n_layer = 8, n_neuron = 512, learning_rate = 0.001 and optimizer = Adam. The result was accuracy of 87.00%, precision of 87.34%, recall of 87.49%, and f1-score of 86.98%.

Table 8 shows the best result of MobileNet-v2 with Epoch = 1000, n_layer = 18, n_neuron = 512, learning_rate = 0.001 and optimizer = Adam. The result was accuracy of 95.50%, precision of 95.53%, recall of 95.68%, and f1-score of 95.57%.

Table 8. Report MobileNet-v2.

Parameter		Class	Precision	Recall	F1-score
Epoch	1000	Overripe (0)	89.80	93.62	91.67
n_layer	18	Semi-ripe (1)	100	100	100
n_neuron	512	Unripe (2)	94.23	89.09	91.59
learning_rate	0.001	Ripe (3)	98.08	100	99.03
optimizer	Adam	Accuracy = 95.50			

Table 9. Report MobileNet-v3.

Parameter		Class	Precision	Recall	F1-score
Epoch	1000	Overripe (0)	79.59	82.98	81.25
n_layer	13	Semi-ripe (1)	95.65	93.62	94.62
n_neuron	1024	Unripe (2)	79.63	78.18	78.90
learning_rate	0.001	Ripe (3)	96.08	96.08	96.08
optimizer	Adam	Accuracy = 87.50			

Table 9 shows the best result of MobileNet-v2 with Epoch = 1000, n_layer = 13, n_neuron = 1024, learning_rate = 0.001 and optimizer = Adam. The result was accuracy of 87.50%, precision of 87.74%, recall of 87.71%, and f1-score of 87.71%.

5 Conclusion

In conclusion, this research evaluated the performance of models including GoogleNet (Inception-v2, Inception-v3), VGGNet (VGGNet16, VGGNet19), ResNet (ResNet-50, ResNet-101, ResNet-152), and MobileNet (MobileNet-v2, MobileNet-v3) for classifying the ripeness level of Monthong durian. The system utilized deep transfer learning, partitioning the dataset into 70% training, 10% validation, and 20% testing sets. All models achieved strong performance, with MobileNet-v2 having the highest accuracy of 95.50%, precision of 95.53%, recall of 95.68% and f1-score of 95.57%.

While these results are promising, there are several recommendations for future work. First, expanding the dataset to include a wider range of Monthong durian cultivars, harvest times, cultivation techniques, and geographical origins would enhance model generalization and improve performance. Additionally, comparing different fine-tuning approaches and hyperparameters for the deep transfer learning models could provide insight into optimizing accuracy. Testing different learning rates, for example, would illuminate their effect on ripeness classification effectiveness.

Moreover, exploring more recent, specialized transfer learning models tailored for agricultural and food-related tasks could boost performance. By prioritizing these avenues, researchers may further extend the detectable ripeness range of Monthong durian using deep transfer learning. This could ultimately facilitate the development of more accurate, robust, and applicable models for evaluating end-user satisfaction. In summary, the high performance achieved thus far under-scores the potential of deep transfer learning for precisely determining Monthong durian ripeness to meet consumer expectations.

References

1. Boonthanakorn, J., et al.: Quality preservation of fresh-cut durian cv. Monthong using micro-perforated PET/PE films. Food Packag. Shelf Life **23**, 100452 (2020)
2. Ketsa, S., Wisutiamonkul, A., Palapol, Y., Paull, R.E.: The Durian: Botany, Horticulture, and Utilization, 47th end. Horticultural Reviews, John Wiley & Sons, Inc., New York (2020)
3. Wiangsamut, B., Wiangsamut, M.E.L.: Assessment of natural fruit ripening and fruit quality of three elite durian cultivars for overland export. Trends Sci. **20**(5), 4647–4652 (2023)
4. Airriess, C.: Trade, tourism and the environment in Asia's durian boom. Focus. Geogr. **63**(1), 6–22 (2020)
5. Treeamnuk, T., Chuenatsadongkot, T., Treeamnuk, K.: Evaluation of Monthong durian maturity using color value from image analysis. Thai Soc. Agric. Eng. J. **24**(2), 38–47 (2018)
6. LeCun, Y., et al.: Backpropagation applied to handwritten zip code recognition. Neural Comput. **1**(4), 451–551 (1989)
7. LeCun, Y., et al.: Gradient-based learning applied to document recognition. Proc. IEEE **86**(11), 2278–2324 (1989)
8. Cahall, D.E.Y., et al.: Inception modules enhance brain tumor segmentation. Front. Comput. Neurosci. **13**(44), 1–8 (2019)
9. Zhiqi, Y.: Face recognition based on improved VGGNET convolutional neural network. In: 2021 IEEE 5th Advanced Information Technology, Electronic and Automation Control Conference (IAEAC), pp. 2530–2533. IEEE, Chongqing China (2021)
10. Peng, S., et al.: More trainable inception-ResNet for face recognition. Neurocomputing **411**, 9–19 (2020)
11. Phan, H., et al.: Binarizing mobilenet via evolution-based searching. In: Proceedings of the IEEE/CVF Conference on Computer Vision and Pattern Recognition, pp. 13420–13429. IEEE, Piscataway New Jersey (2020)
12. Lim, M.G., Chuah, J.H.: Durian types recognition using deep learning techniques. In: 2018 9th IEEE Control and System Graduate Research Colloquium (ICSGRC), pp. 183–187. IEEE, Malaysia (2018)
13. Kharamat, W., Wongsaisuwan, M., Wattanamongkhol, N.: Durian ripeness classification from the knocking sounds using convolutional neural network. In: 2020 8th International Electrical Engineering Congress (iEECON), pp. 1–4. IEEE, Chiang Mai Thailand (2020)
14. Muresan, H.: Fruit recognition from images using deep. arXiv 10(1), 26–42 (2018)
15. Image classification in Galaxy with Fruit 360 dataset 2018. https://training.galaxyproject.org/training-material/topics/statistics/tutorials/fruit_360/tutorial.html. Accessed 7 Jan 2024
16. Hongboonmee, N., Jantawong, N.: Apply of deep learning techniques to measure the sweetness level of watermelon via smartphone. J. Inf. Sci. Technol. **10**(1), 59–69 (2016)
17. Chaikaew, A., et al.: Convolutional neural network for pineapple ripeness classification machine. In: 2019 16th International Conference on Electrical Engineering/Electronics, Computer, Telecommunications and Information Technology (ECTI-CON), pp. 373–376. IEEE, Pattaya Thailand (2019)
18. Gallenero, J.A., Villaverde, J.: Identification of durian leaf disease using convolutional neural network. In: 2023 15th International Conference on Computer and Automation Engineering (ICCAE), pp. 172–177. IEEE, Sydney Australia (2023)

19. Alem, A., Kumar, S.: Deep learning models performance evaluations for remote sensed image classification. IEEE Access **10**, 111784–111793 (2022)
20. Amin, M., et al.: ECG-based driver's stress detection using deep transfer learning and fuzzy logic approaches. IEEE Access **10**, 29788–29809 (2022)
21. Aranha, G.D.A., Fernandes, R.A.S., Morales, P.H.A.: Deep transfer learning strategy to diagnose eye-related conditions and diseases: an approach based on low-quality fundus images. IEEE Access **11**, 37403–37411 (2023)
22. Miraei Ashtiani, S.-H., et al.: Detection of mulberry ripeness stages using deep learning models. IEEE Access **9**, 100380–100394 (2021)
23. Maschler, B., Weyrich, M.: Deep transfer learning for industrial automation: a review and discussion of new techniques for data-driven machine learning. IEEE Ind. Electron. Mag. **15**(2), 65–75 (2021)

Author Index

P. Meesad et al. (Eds.): IC2IT 2024, LNNS 973, p. 163, 2024.
https://doi.org/10.1007/978-3-031-58561-6

Printed in the United States
by Baker & Taylor Publisher Services